W9-DHR-622

# PRAISE FOR *THE BUTTON*

"Every citizen should read this book—a clear account of our history with nuclear weapons, the continuing risks of their use through human error or cyberattacks, and the authors' recommended steps to create a safer future. It's an assault on the complacency of 'nobody would be crazy enough to unleash these' thinking."

**—President Bill Clinton**

"Nuclear weapons pose an urgent, existential threat to mankind. In *The Button*, William J. Perry and Tom Z. Collina describe the grave but largely forgotten danger we now face—and how we can greatly reduce it. Perry has more than half a century of experience dealing with nuclear issues at the highest levels of government. Collina is one of our most brilliant and incisive nuclear weapon experts. Together they have devised a number of simple ways to replace the current madness with common sense. Today a nuclear catastrophe could occur instantaneously, at any moment, without any warning, with a lasting impact too terrible for words. That's why *The Button* is one of the most important books of 2020."

**—Eric Schlosser, Pulitzer Prize finalist and author of *Command and Control***

"Bill Perry, one of the wisest and most effective secretaries of defense ever to serve, has coauthored with Tom Collina another provocative, must-read book. *The Button: The New Nuclear Arms Race and Presidential Power from Truman to Trump* brings to life the nuclear dangers and dilemmas of the present day and makes a compelling case for several pragmatic changes to US nuclear weapons policy that would meaningfully reduce the risk of nuclear miscalculation. Whether you consider yourself a hawk or a dove, an expert or an interested citizen, *The Button* is required reading for anyone who wants to ensure we avoid nuclear war in an era of great power competition."

**—Michèle Flournoy, former US Undersecretary of Defense for Policy**

"At a time when the power to destroy our world rests in the hands of men like Donald Trump and Vladimir Putin, *The Button* is a powerful and urgent reminder that the risk of nuclear war is far too great. More than that, it offers a smart, comprehensive, well-argued case for what we can do to pursue a safer and more peaceful world."

**—Ben Rhodes,**
**Deputy National Security Advisor to President Obama**

"The risk of accidental nuclear war is increasing, and through *The Button* Perry and Collina give an insightful account of the dangers of nuclear weapons, how fragile the current nuclear launch system is, and most important—what you can do about it. This book will make you realize that no one person should have the sole authority to end the world and there is an urgent need to move to prohibit and eliminate nuclear weapons."

**—Beatrice Fihn,**
**recipient of the 2017 Nobel Peace Prize**

"No one knows how to prevent nuclear war better than Bill Perry. He had a front row seat to the arms race and the wisdom to back away from the brink. In this must-read book, Perry and Collina tap into a powerful insight—that we have been focused on the wrong nuclear threat. They chart a new and compelling course to keep us safe from blundering into atomic destruction."

**—Wendy Sherman,**
**former Undersecretary of State for Political Affairs**

"Bill Perry and Tom Collina give some clear thinking about the dangers of nuclear weapons and how to reduce these dangers. This deserves top attention at capitals around the world."

**—George Shultz, former US Secretary of State**

"Former secretary of defense Bill Perry, the strategist behind most of the US military advantage today, and nuclear policy expert Tom Collina have defined the growing nuclear risks that could determine whether our nation—and indeed the world—survives. In our cyber world with nine nuclear weapons states, the risk of blunder, mistake, or false warnings greatly exceed the risk of a premeditated nuclear attack. Our strategies reflect old thinking, leaving us exposed to grave and unnecessary dangers. *The Button* is a must-read book for leaders and citizens and underscores the urgent need for new thinking and wise, rational leadership on the most important issues facing the world."

**—Sam Nunn, former US Senator**

"American nuclear policy is broken. We are drifting toward catastrophe with a new arms race, new weapons, and new war-fighting doctrines pulled from the darkest days of the Cold War. In *The Button*, two top experts, William Perry and Tom Collina, detail the problems and lay out exactly the dramatic shift we need to pull us back from the brink. There is no more important issue than preventing nuclear war, and no more important time than now. If you read only one national security book this year, read this one."

**—Joe Cirincione,**
**former president of the Ploughshares Fund**

"*The Button* is about presidents and their apocalyptic power; about the stark danger America faces from its nuclear overkill. In chilling detail, Perry and Collina recount the several instances when Russia and the United States almost fired on one another—by mistake. In mere minutes, the president of the United States—on his own—can launch thousands of nuclear missiles solely because of warning messages from a computer. This book is a wake-up call: timely and profoundly important."

**—Jerry Brown, former Governor, California**

# THE
# BUTTON

## ALSO BY WILLIAM J. PERRY

*My Journey at the Nuclear Brink* (Stanford University Press, 2015)

*Preventive Defense: A New Security Strategy for America* (by William J. Perry and Ashton B. Carter; Brookings Institution Press, 1999)

# THE

# BUTTON

## THE NEW NUCLEAR ARMS RACE
## AND PRESIDENTIAL POWER FROM
## TRUMAN TO TRUMP

### WILLIAM J. PERRY
### AND TOM Z. COLLINA

BenBella Books, Inc.
Dallas, TX

The views expressed in this book are those of the authors and do not necessarily express policy or position of the Department of Defense or the US Government. The public release clearance of this book by the Department of Defense does not imply Department of Defense endorsement or factual accuracy of the material.

BenBella Books, Inc.
10440 N. Central Expressway, Suite 800
Dallas, TX 75231
www.benbellabooks.com | Send feedback to feedback@benbellabooks.com

*BenBella* is a federally registered trademark.

Printed in the United States of America
10 9 8 7 6 5 4 3 2 1

Library of Congress Control Number: 2019059814
ISBN 9781948836999 (hard cover) | ISBN 9781950665181 (electronic)

Editing by Stephanie Gorton
Copyediting by Karen Levy
Proofreading by Marissa Wold Uhrina and Greg Teague
Indexing by WordCo Indexing Services, Inc.
Text design by Katie Hollister
Cover design by Faceout Studio, Tim Green
Cover photo © Shutterstock / jejim (eagle), caesart (texture), and Lack-O'Keen (nuclear symbol)
Photos on pages 145–148 courtesy of George Washington University National Security Archives
  (www.nsarchive.org), on page 160 © David Hume Kennerly/Getty Images (top) and Olivier
  Doulier—Pool/Getty Images (bottom)
Printed by Lake Book Manufacturing

Distributed to the trade by Two Rivers Distribution, an Ingram brand
www.tworiversdistribution.com

*To Sara, Jared, Natalie, Lu, and my entire family,*
*you make this a world worth saving.*
—Tom Z. Collina


*To my twelve grandchildren, who give me the best of all reasons*
*to continue working to avert a nuclear catastrophe.*
—William J. Perry

# CONTENTS

# CONTENTS

# PREFACE

# "YOUR SHOT,
# MR. PRESIDENT"[1]

I t's a cool fall day on the golf course, a stiff breeze moving puffy white clouds through the sky. The afternoon sun angles through the trees, putting everything in sharp, clear contrast. The world in high definition. The president of the United States is considering his next shot. His ball lies in the rough, about a hundred yards from the green, and so deep in the weeds that he can barely see it. He doesn't have time for this, yet he is badly in need of distraction. His domestic political problems are boiling over, something to do with tax evasion years ago, and Congress is moving to start impeachment hearings. If that weren't bad enough, Russia is moving a menacing military force to its border with Belarus, where a pro-Western government is talking about joining NATO. The Russian president said that morning that Belarus would turn to the West "over my dead body" and reminded Washington that when Ukraine played this game in 2014, "it came to a bad end."

Just a month before what he hoped would be his reelection day, the US president is determined to do three things: distract public attention from Congress's impeachment shenanigans, save Belarus and make Democrats look weak for having "lost" Ukraine, and, above all, look "presidential."

"Your shot, Mr. President."

The president peers into the weeds, trying to imagine how to get his ball onto the green. A cell phone ring breaks his trance. His national security advisor answers, stiffens, and rattles off a series of words. "NSA Ellen Banks. Charlie. Delta. Three. Three. Niner. Seven."

"What's going on?" the president demands, annoyed that his game is being interrupted.

"Early warning satellites detected multiple Russian launches," Banks says, her voice with an edge the president has not heard before. "Two hundred missiles in flight."

"What?! There must be a mistake. The Russians can't be that crazy. Nuclear war over Belarus?"

"New intel says Russian troops are pouring over the border toward Minsk. Shots fired."

"Mr. President, I have STRATCOM on the line. Putting on speaker."

"Mr. President, this is General Bradley. I am sorry to bother you, but my screens now show four hundred Russian ICBMs heading for the US mainland, striking in about ten minutes."

"My God. Do we have confirmation? Could it be a false alarm?"

"London and Alaska already confirmed. No indications of false alarm."

"Could it be a computer hack?"

"Not possible. Every indication is that we are under attack, sir."

"Mr. President, I recommend scrambling our strategic bombers immediately and initiating launch procedures. And let's get on the chopper to get you to a safe location."

"Do it," the president says as he rushes onto the helicopter. "Major, open the football."

A chill goes down the president's spine as he says these words for the first time. Since the moment he took the oath of office almost four years ago, a senior military aide has followed the president everywhere he goes carrying the "football," a briefcase that contains a secure phone, identification codes, nuclear attack options, and anything else the president might need to launch a nuclear strike.

"Okay, team, game time. What are my options?"

"Major attack option three takes out all of Russia's military installations and limits civilian casualties," the national security advisor says. "But it may not neutralize the entire military."

"Mr. President, I strongly advise major attack option one. Full counter-strike," implores the defense secretary. "Anything less invites a second wave, killing even more Americans."

"Mr. President, I concur," says the national security advisor. "And we are running out of time. If you don't launch before the Russian missiles hit, they could take out our silos, severely limiting our response. We would lose an entire leg of the triad."

The science advisor speaks up: "Hold on. There has to be a better option than responding in kind to an all-out attack. The Russian attack alone will bring on nuclear winter, killing almost everyone on the planet, including the Russians. Why pile on? And don't forget, this could still be a false alarm. They have happened before, just like this. Do you want to go down in history as the president who started World War Three by mistake?"

"She's right, Mr. President," says Banks. "As much as I hate to say it, it would be better to wait for the Russian missiles to land, so at least we know it's a real attack. That way we know what we are dealing with. And if it is a real attack, we could still retaliate with our submarines at sea."

"Come on, get a grip, people! Nuclear winter?" the defense secretary barks dismissively. "It's just a theory. Science fiction. We have been gaming and planning for this moment for decades. By the book. There's only one way to fulfill your constitutional obligations, Mr. President. It's major attack option one."

The president looks up. "Does anyone disagree with that?"

The only sound is the whump-whump-whump of the chopper blades.

"Mr. President," the national security advisor speaks urgently. "Incoming missiles will land in about four minutes. But we also have new information. Not all of our radars are seeing the attack. There is disagreement among some of our sensors."

"It's a false alarm," says the science advisor. "We must not respond until we are sure."

"If you don't launch now it will be too late."

"I was elected to lead. I will not sit by and do nothing in the face of nuclear attack," the president huffs. "I order major attack option one. Get the War Room on the line."

"General Nelson here, Mr. President. Ready to authenticate the nuclear launch codes on your call."

The national security advisor interrupts. "Wait! Mr. President, we just received an urgent telex on the hot line from Moscow. They have seen our SAC bomber crews scrambling. They are concerned and are reassuring you that there is no nuclear attack underway."

"Normally I would not trust that," says the science advisor, "but please realize that we have received no confirmation from human intel that missiles are in the air. All indications are from computer networks that we know are fallible. This doesn't feel right. We can't launch a nuclear attack unless we're sure."

The defense secretary's tone shifts. "I just got a text from my Russian counterpart, General Yukov. He says there's no attack. I am skeptical, but I trust this man."

"Two minutes to impact, Mr. President."

The president waves his arms. "You're all going soft. Moscow's sending troops into Belarus. They're investing billions in new nukes. What more do you need to know? They're lying about the attack—any fool can see that! Why do we have these weapons if we're not willing to use them? I've already decided to launch, and I don't need anyone's approval to do it. Not Congress, not the Pentagon, not anyone. Let's go. General Nelson, you still on the line?"

"Yes, sir."

"I want major attack option one, now!"

"Okay, Mr. President, are you ready to authenticate? Delta, Zulu, Six, Echo, Foxtrot, Five."

The national security advisor hands the president a small card, known as the "biscuit."

"Zulu. Echo. Six. Two. Five. Eight. Repeat. Zulu. Echo. Six. Two. Five. Eight."

"Copy, Mr. President. Godspeed, sir."

"Let the birds fly," the president declares.

His orders confirmed, General Nelson sends encrypted messages from the National Military Command Center in the Pentagon to all ICBM (intercontinental ballistic missile) launch control stations, Trident submarines at sea, and strategic bomber bases. Within one minute, four-hundred US ICBMs lift off from their silos, and within ten minutes submarine-launched missiles join the fray—almost a thousand warheads in all. Once launched, these weapons cannot be recalled.

And then—nothing. Just nothing. No explosions, no mushroom clouds, just the din of the chopper blades. After five endless minutes, it becomes clear that there is no incoming Russian attack. The president and his staff begin to process the awful implications.

"What happened?" the president yells.

"STRATCOM reports it was a false alarm."

"Then call off the attack!"

"Mr. President, the missiles have already been launched."

"Then call them back!"

"That's not possible, sir."

"What do you mean? Can't we blow them up or something?"

"I'm afraid not, sir. We decided long ago that a remote self-destruct mechanism could be hacked by the enemy to destroy our weapons. Better safe than sorry."

"Well, can we alert the Russians, so they know this was all a mistake, no hard feelings? Maybe they can shoot them down?"

"We can try, sir, but I'm not sure they'll believe us, given the investment we've made in building new weapons. And their missile defense system only protects Moscow and, anyway, we regard it as ineffective."

"So, what can we do now?"

"Tell the Russians it's a false alarm. Hope for the best. And get ready for their response."

About ten minutes later, the watch officer on duty in Russia sees the indications of a massive attack. Obviously, the Americans did not believe their message of reassurance. The watch officer alerts his supervisor, who alerts her supervisor, who alerts the Russian president, who just received an urgent message from the US president that, despite what he may be seeing and hearing, there is no US attack underway. "Nice try," says the Russian president.

There was no Russian launch, but now there will be. The Russian president orders a massive counterstrike, one thousand warheads in all. Within thirty minutes, it's all over.

The damage is devastating. All major cities in both countries are obliterated, and many tens of millions are dead instantly from the blasts and the radiation. They are the lucky ones. Hundreds of millions are seriously wounded. But there are no hospitals left to go to, no doctors to treat the survivors. Civil society collapses; mass chaos ensues. And, as it turns out, the American science advisor was right about nuclear winter. Smoke and soot from burning cities soon encircle the globe, blocking sunlight from reaching the earth for years.

No sunlight, no warmth . . . no food. Nearly all humans—seven billion people—starve within a few years. Even the southern hemisphere, which avoided direct nuclear attack, will not survive nuclear famine.

Nations that did not possess the bomb knew that a nuclear war, even one that did not target them directly, would drag them down with it. These nations had been agitating for decades—through the United Nations and other fora—for nuclear disarmament. Finally, in 2017, the UN passed a treaty to ban the possession and use of nuclear weapons. But none of the states with nuclear weapons signed it. Even the United States, which at the time actively supported the elimination of nuclear weapons, did not support the ban treaty. The previous US administration said it was not the right approach.

The initial Russian nuclear "attack" was indeed a false alarm. An unknown adversary had hacked into the computers at US Strategic Command and manipulated the system to show a massive missile launch.

The US president's helicopter never made it to a secure location. En route, a Russian warhead detonated close to the chopper and engulfed it in flames. The president's last act was to pick up his cell phone and tweet that Russia had just started a nuclear war, and that he had no choice but to retaliate. There was no cell service.

Authors' note: Everything in this fictional account could happen and is consistent with US policies. For example:

1. Under current US policy, the president has the authority to launch nuclear weapons first and is not limited to retaliation; to launch nuclear weapons under warning of attack, rather than wait for evidence of attack; and to launch nuclear weapons on his or her sole order.
2. US nuclear weapons are kept on high alert and, in the case of land-based missiles, can be launched within minutes.
3. Once launched, nuclear-armed ballistic missiles cannot be recalled.
4. Both the United States and Russia are investing trillions of dollars in new nuclear weapons.
5. False alarms have happened multiple times and can happen again. For example, in 1980, a false alarm was reported to the national security advisor and was almost reported up to President Jimmy Carter as a real attack but was luckily identified in time.
6. Recent Pentagon reports have found that, as a result of cyberattacks, the president could be faced with false warnings of attack or lose the ability to control nuclear weapons.

The threat of nuclear annihilation feels distant to many people, but many aspects of the scenario we just described have happened and could in fact happen today. The president possesses sole authority to launch America's arsenal of nuclear weapons. With this book, we illuminate the profound history of this dangerous policy and the insights necessary to pressure our government to institute responsible changes to avoid the tragic consequences of nuclear weapons.

We, Tom Collina and Bill Perry, come to this topic from starkly different backgrounds. I (Tom) have followed Bill Perry and his work since his years in the Clinton administration, but it was his 2015 memoir, *My Journey at the Nuclear Brink*, that brought us together. Bill began a high-profile effort to educate the public on the continuing dangers of nuclear war, a drum I've been beating for thirty years—but from outside of government. Bill Perry is the ultimate insider, holding a front-row seat to the history of nuclear weaponry. His career in the military first began as a young occupying soldier in Japan right after World War II, where he came to understand the devastation wrought by atomic bombs. He worked as a technical consultant to the CIA during the Cuban Missile Crisis, and in senior Pentagon posts for President Carter during the Cold War, ultimately serving as secretary of defense under President Bill Clinton. As former California governor Jerry Brown has said: "I know of no person who understands the science and politics of modern weaponry better than William J. Perry."

What makes Bill unique is that he has come full circle. During the Cold War, he supported doomsday nuclear weapons like the MX missile, the B-2 bomber, and the air-launched cruise missile. But even then, he was also a strong supporter of United States–Russian arms control treaties like the Strategic Arms Reduction Treaty (START) and the nuclear test ban. This gives him tremendous credibility when he argues against the new nuclear weapons being proposed by President Donald Trump. Today, members of Congress who want to craft a progressive nuclear policy seek out his wise counsel because he has been there—he walked up to the nuclear brink, looked into the abyss, and had the wisdom to turn away.

As Bill wrote in 2016: "Russia and the United States have already been through one nuclear arms race. We spent trillions of dollars and took incredible risks in a misguided quest for security. I had a front-row seat to this. Once was enough. This time, we must show wisdom and restraint. Indeed, Washington and Moscow both stand to benefit by scaling back new programs before it is too late. There is only one way to win an arms race: Refuse to run."

The key issue Bill and I came together on was the risk of stumbling into nuclear war. We both realized, from our separate vantage points, that the threat of Russia intentionally launching a nuclear attack against the United States was vanishingly small. Smaller, indeed, than the risk that we might start a nuclear war by mistake. And once you make that mental shift—that the real threat is blundering into war—you come to see existing nuclear policy as very, very dangerous. The key facts—that the president can order nuclear war on his own authority, that US weapons can be used first, that US weapons are on hair-trigger alert and ready for use, that weapons are susceptible to cyberattack, and that we have hundreds of vulnerable land-based missiles—all increase the danger of catastrophe by accident.

Yet the military establishment does not see this danger. It thinks we should keep our Cold War nuclear policies and double down on them by investing $2 trillion to rebuild the same system and policies that we had during the Cold War—even though that system increases the danger of the primary threat: stumbling into nuclear war.

Today's military leaders have learned the wrong lessons from the Cold War and are keeping the most dangerous aspects of US nuclear policy when they no longer need to. Add to the equation an impulsive President Trump with his finger on the button and you have a combustible mix. I am thrilled and honored to be taking on this crucial issue with Bill. He is the quintessential nuclear expert of his time. His extraordinary life experiences, wisdom, and character put him in a league by himself.

*Bill here.* While my conflicts have been with presidents and Pentagon offices, Tom's battles are often fought on the steps of Congress for everyone to see. Tom realized early on that engaging the public in critical debates, even about subjects as complex as nuclear proliferation, is key to our efforts in changing policy. As director of policy at Ploughshares Fund and for other nongovernmental groups, Tom has worked outside of government for thirty years to promote reductions in nuclear arsenals, stop their spread to other countries, and limit nuclear risks. After decades of work, I have come to believe that when the public understands nuclear danger, only then will they demand changes in nuclear policy. I value Tom's long experience engaging the public in these complicated issues and his deep, practical knowledge of the political landscape. Our previous collaborations have produced effective and insightful work and built a mutual trust. I am honored to coauthor this book with him and to share strategies for building a saner, safer world. While I am in the sunset of my work on nuclear danger, Tom is part of the next generation of leaders so badly needed to carry this work forward. Together, we have a comprehensive view of the dangers posed by nuclear weapons, the forces that keep the nuclear establishment alive, and how to overcome those forces so the United States might chart a different course.

But how did a former defense secretary who helped create nuclear weaponry and a rabble-rousing policy wonk get on the same page?

Our paths did not cross even remotely until the early years of the Clinton administration, when we were both working to stop US nuclear testing and create a global ban. Bill was deputy and then secretary of defense, and Tom was director of the Global Security Program at the Union of Concerned Scientists. After years of effort both inside and outside of government, we each in our own ways helped make it possible for President Clinton to sign the Comprehensive Test Ban Treaty (CTBT) in 1996.

Bill left government soon thereafter and settled in as a go-to person for the media and members of Congress to get up to speed on the latest nuclear dangers. Bill started traveling from Palo Alto, California, to Washington, DC, regularly, and Tom began to help Bill (and his daughter and project associate Robin Perry) organize briefings for reporters, congressional staff, and one-on-one meetings with senators and representatives.

Soon, we started joint writing projects. Bill wrote an essay for a 2016 Ploughshares Fund report, *Ten Big Nuclear Ideas for the Next President*, calling for phasing out US land-based ballistic missiles—a radical idea for a former secretary of defense then *or* now. After the 2016 election, we coauthored op-eds in opposition to President Trump's proposal for new "low-yield" nuclear warheads. In 2018, we both sensed rising public concern about President Trump's sole authority to launch nuclear weapons. So, we decided to write this book, a history of the nuclear button, based on the insights of the people most experienced with this dangerous remnant from the Cold War.

In addition to our own voices, we have included those of many nuclear experts in *The Button*. These include witnesses and makers of nuclear history, such as former president Bill Clinton, former defense secretary Jim Mattis, 2017 Nobel Peace Prize winner Beatrice Fihn, and current chairman of the House Armed Services Committee, representative Adam Smith. We are grateful for the time and insights these and other experts have contributed to our effort.

It can be a challenge to write a book with two authors on different coasts, so the reader may notice as the narrative shifts from Tom's voice to Bill's, and back again. We wanted to include personal stories and anecdotes, so at times we shift from "we" to "I" to make it clear who is telling the story. However, our point of view is singular based on our combined one hundred years of experience with nuclear weapons. There is no more important story to tell, and no better time to tell it.

— PART I —

# THE WRONG THREAT

# CHAPTER 1

# THE PRESIDENT'S WEAPONS

*I can go back into my office and pick up the telephone and in
25 minutes 70 million people will be dead.*
—President Richard Nixon[1]

Some believe that President Donald Trump, given his temperament, should be the *last* person entrusted with the authority to launch nuclear weapons. He is, in fact, the *only* one who is.

On January 20, 2017, President Trump got the keys to the US nuclear arsenal, the deadliest killing machine ever created. The president said it was "a very sobering moment, yes. It's very, very scary, in a sense."[2] While that statement might sound reassuring, he also said, on a separate occasion, "If we have nuclear weapons, why can't we use them?"[3]

As long as President Trump is in the White House, he will have the same frightening power as all presidents since Harry Truman, whose two terms ran from 1945 to 1953. Within minutes, with just one phone call, President Trump could unleash up to a thousand nuclear weapons, each one many times more powerful than the Hiroshima bomb. It would be the end of civilization. Short of mutiny, no one can stop him. Once launched, the missiles cannot be recalled. For President Trump, starting nuclear war is about as easy as sending a tweet.

"Having some understanding of the levers that a president can exercise, I worry about, frankly, the access to the nuclear codes," former director of National Intelligence James Clapper said in 2017. If "in a fit of pique [Trump] decides to do something about Kim Jong-un, there's actually very little to stop him. The whole system is built to ensure rapid response if necessary. So there's very little in the way of controls over exercising a nuclear option, which is pretty damn scary."[4]

We see this as a profoundly dangerous situation, yet the American public is largely unaware of this reality. Only about 25 percent of the public knows that the president has sole nuclear authority.[5] Many more (about 44 percent) think the president must get congressional approval; others think there must be consultation with the defense secretary (20 percent) or the Joint Chiefs of Staff (12 percent). In fact, while the president can *choose* to consult with advisors, he is not *required* to do so.

Putting this much military power in the hands of one person goes against everything for which the American Constitution stands. As historian Michael Beschloss writes in *Presidents of War*, "The Founders created a Constitution that gave Congress the sole power to declare war, and divided the responsibility to wage war between the executive and legislative branches. As Congressman Abraham Lincoln wrote to his friend William Herndon in 1848, the early Americans resolved that '*no one man* [italics Lincoln's] should hold the power' to take the nation to war."[6]

America's Founders tried to constrain presidential power, but they could not have anticipated the discovery of the nuclear bomb and how it would change the nature of the presidency. Today, the Founders would be shocked to learn that the president alone has the authority to launch nuclear weapons that could destroy the human race.

"There is nothing stopping a nuclear war except for Donald Trump's brain," said Ben Rhodes, who served as deputy national security advisor to President Barack Obama. "That should be concerning to people. There should at least be some check, some process, some chain of command, some congressional notification, some form of break in which people can stop and consider even for just a brief period of time: Do we really want to do this? Especially [with] first use. I don't know how anybody can sit here and think that it's rational that a person with the manner of Donald Trump—and we could elect someone like that again—has completely, with their own discretion, the capacity to destroy life on earth."[7]

We spoke with President Bill Clinton for this book, a former possessor of the nuclear button who is now open to rethinking it. He told us, "If we could construct a constitutional deliberative process [where], before the president can do this, you must hear from one, or two, or three people or agencies that guarantee perspective, experience, and raw knowledge, and reminding the president of why nobody's dropped a nuclear weapon in a very long time, I think that would be a good thing."[8]

During his presidency, when advisors pushed Clinton to take immediate military action to avoid looking "weak," his response was, "Can I kill them tomorrow? Because if I can, we're not weak. I certainly can't bring them back to life tomorrow. So, what do you say we try one more day to work this out? . . . If you drop a nuclear weapon, God forbid ever, most of the casualties will be completely innocent collateral damage."

"The main thing," Clinton said about creating time for consultation, "is it gives everybody a chance to take a deep breath and make darn sure they want to risk the whole future of humanity on this roll of the dice."

There are many control systems in place to prevent nuclear weapons from being launched unintentionally or by an unauthorized person. But currently there is no way to prevent a determined president from starting a nuclear war. We strongly believe that the risks of having nuclear weapons ready to launch within minutes, on the president's sole authority, outweigh any perceived benefits. This system is unconstitutional, dangerous, outdated, and unnecessary.

## IT'S NOT ALL ABOUT TRUMP

President Trump's temperament has brought attention to the awesome power that he holds, but he is not the first president to raise concerns about sole authority—and he is unlikely to be the last. As President Richard Nixon famously boasted at the height of the Watergate scandal in 1974, "I can go back into my office and pick up the telephone and in 25 minutes 70 million people will be dead."

Former defense secretary James Schlesinger feared that President Nixon, who seemed depressed and was drinking heavily at the end of his presidency, might go off the deep end. Senator Alan Cranston phoned Schlesinger, warning about "the need for keeping a berserk president from plunging us into a holocaust." Schlesinger told military commanders that if the president ordered a nuclear launch, they should check with him or Secretary of State Henry Kissinger first. But for all of Schlesinger's very real concerns, there was not much he could do. There was no guarantee that the commanders would have honored his request, which had no legal basis.[9]

In addition to Nixon, President John F. Kennedy made heavy use of pain medications, which could have clouded his thinking at a time of crisis.[10] And President Ronald Reagan, who was formally diagnosed with Alzheimer's five years after leaving office, was reported to have started showing signs of the disease while still in the White House.[11] Some presidents may appear or claim to be rational decision makers, but in fact they are not. They are all too human.

A US president can launch nuclear weapons unilaterally, at any time, without any provocation. Given the fallible nature of humans, certainly including presidents, this is a grave concern. But we worry even more about what a president might do in a crisis situation, where emotions run high and decision time is short.

Take, for example, the false alarm scenario described in the preface. No president should have to decide the fate of the world in just minutes based on incomplete information. Yet current US policy—the option to shoot first, with weapons on high alert, under pressure to launch on warning of attack or lose weapons on the ground—could force the president into a dangerous corner. In the words of Gen. George Lee Butler, who commanded US nuclear forces at the end of the Cold War, American nuclear policy is "structured to drive the president invariably toward a decision to launch under attack."[12]

And there are other ways presidents can stumble into nuclear war. In summer 2017, President Trump and North Korea leader Kim Jong-un were locked in a tense political standoff over the North's nuclear weapons. That August, President Trump threatened Pyongyang: "North Korea best not make any more threats to the United States. They will be met with fire and fury like the world has never seen." It is hard to interpret this any other way than as a nuclear threat.

In October 2017, while in Puerto Rico, Trump reportedly pointed to the "nuclear football"—the briefcase used to authorize a nuclear attack—and claimed he could use it on Kim whenever he wanted to. "This is what I have for Kim," Trump said.[13]

Then, in early January 2018, Kim Jong-un said his nuclear launch button was "always on my table." In response, President Trump tweeted: "Will someone from his depleted and food starved regime please inform him that I too have a Nuclear Button, but it is a much bigger & more powerful one than his, and my Button works!"[14]

Soon thereafter, on January 13, the Hawaii Emergency Management Agency broadcast an official message: "Ballistic missile threat inbound to Hawaii. Seek immediate shelter. This is not a drill." Hawaiians waited in terror nearly forty minutes for a nuclear attack that never came. It was a false alarm—the alert was sent as the result of human error—but given the provocative rhetoric on both sides, the attack was all too easy to believe.

This incident only underscores how vulnerable we are to a technical or human error in our warning system, especially if the false alarm comes at a time of heightened political stress. And what could be more stressful than two unpredictable nuclear-armed leaders boasting about their nuclear buttons?

## RISING PUBLIC CONCERN

Soon after Trump's tweet about the size of his nuclear button, a *Washington Post*–ABC News poll found that over half of Americans were concerned that

President Trump might launch a nuclear attack without justification. According to the *Post*, "Overall, 38 percent of Americans trust Trump to handle the authority to order nuclear attacks on other countries, while 60 percent do not. Among those who distrust Trump, almost 9 in 10 are very or somewhat concerned the president might launch an attack."[15]

Once informed about sole authority, nearly 80 percent of Americans are concerned about President Trump having this unchecked power. Among this group, over 70 percent support limiting the president's nuclear authority, which they see as putting too much power in the hands of a single, fallible person.[16]

Congress is taking note as well. In November 2017, after Trump's comments about "fire and fury," the US Senate Foreign Relations Committee held its first hearing in forty-one years on the president's authority to launch nuclear weapons. As Senator Chris Murphy (D-CT), put it, Americans are concerned that President Trump "is so unstable, is so volatile" that he might order a nuclear strike that is "wildly out of step" with our national security interests. "Let's just recognize the exceptional nature of this moment in the discussion that we're having today," he said.[17]

In a rare bipartisan moment, then Senate Foreign Relations Committee chair Bob Corker (R-TN) and then ranking member Ben Cardin (D-MD) agreed that the president has the ability to launch US nuclear weapons on his own authority. "The president has the sole authority to give that order, whether we are responding to a nuclear attack or not," said Corker. "Once that order is given and verified, there is no way to revoke it."[18] Senator Cardin said, "Based on my understanding of the nuclear command-and-control protocol, there are no checks—no checks—on the [p]resident's authority. The system as it is set up today provides the [p]resident with the sole and ultimate authority to use nuclear weapons."[19]

For a brief moment, it seemed that this bipartisan, high-profile attention to unchecked presidential power might motivate Congress to limit this authority. But defenders of the nuclear establishment, a $50-billion-a-year industry employing tens of thousands of people, had no interest in changing the current system to make it conform to the Constitution. So, they sought to manage public anxiety by suggesting that the US system to control the bomb would somehow not allow a president to launch nuclear weapons unless it was truly justified. Gen. C. Robert Kehler, a former head of Strategic Command, sought to reassure the Senate that "the military does not blindly follow orders" and that illegal orders would not be carried out.

However, a president's order to launch would be legal, even if it were profoundly unwise. Military officers are trained to follow orders, not to question them. Those officers who are tempted to question a presidential order of this kind might remember Harold Hering. In 1973, Hering was training to be an Air Force nuclear missile launch officer. He asked his instructors: "How can I be certain that any launch order I receive comes from a sane president?" Rather than provide an answer to Hering's question, the Air Force simply fired him.[20]

A presidential nuclear order is lawful and must be obeyed as long as it comes from the president as commander in chief, acting to protect and defend the nation against an actual or imminent attack. Dr. Peter Feaver from Duke University noted that in the world of military officers, "there is a presumption that the [nuclear launch] orders are legal."

Congressman Ted Lieu (D-CA), who sits on the Foreign Affairs Committee, served in the Air Force, where he met a nuclear missile launch officer. In an interview with us, he recalled that "talking to him, it was very clear that when the order came down, he was launching. There's no way he could tell if it was an illegal order . . . they're trained to simply launch when it happens." That experience convinced him that "just one person could decide essentially the fate of the world for a lot of people. And I always thought that we need to fix that structural problem."[21]

So, let us set the record straight: under current policy, there is no realistic way to stop a determined president from going nuclear. President Trump—like all presidents in the atomic age—has the sole authority to unleash nuclear Armageddon on the world, potentially ending civilization as we know it. Within minutes, the president could unleash the equivalent of more than ten thousand Hiroshima bombs.

One might think that the process of running for president of the United States would weed out individuals who are unfit for this awesome responsibility. As Gen. Maxwell D. Taylor, a former chairman of the Joint Chiefs of

Staff, wrote, "As to those dangers arising from an irrational American president, the only protection is not to elect one."[22]

But what if we do? What if the ultimate control mechanism—the election process—fails us? The potential consequences are too great to just throw up our hands and hope for the best. In the spirit of the American Constitution, we need some real checks and balances on presidential power, regardless of whom we elect.

## LIMITING SOLE AUTHORITY

Congress should be involved with—and even control—the decision to use nuclear weapons in cases other than retaliation. After all, the US Constitution gives Congress, not the president, the power to declare war. The first use of nuclear weapons is the ultimate declaration of war.

Some might say that the Founders could not have foreseen the bomb and thus we cannot guess at what their intentions might have been. Even so, they clearly sought to avoid placing so much authority in one person. As Dr. Kennette Benedict, former director of the Bulletin of the Atomic Scientists, argues, such power completely contradicts the constitutional checks and balances that the Founders created. "We have no voice in the most significant decision the United States government can make—whether to destroy another society with weapons of mass destruction."[23]

When President Dwight Eisenhower was considering the use of the bomb in the 1950s, he acknowledged that unilateral action would be unconstitutional. "If congressional authorization were not obtained, there would be logical grounds for impeachment. Whatever we do must be done in a constitutional manner," the president said.[24] But since then, Congress's role in declaring war has been undermined considerably.

But now members of Congress are reasserting their constitutional authority. A longtime champion of reducing nuclear risks, Senator Ed Markey (D-MA) has partnered with Representative Ted Lieu to introduce legislation that would prohibit the president from initiating the first use of nuclear weapons without the approval of Congress. According to recent polling, two-thirds of Americans, including 60 percent of Republicans, support limiting presidential sole authority in this way.[25]

"Trump's brand is to be unpredictable and rash, which is exactly what you don't want the person who possesses the nuclear football to be," said Representative Lieu. And as Senator Markey said, "Neither President Trump, nor any other president, should be allowed to use nuclear weapons except in response to a nuclear attack. By restricting the first use of nuclear weapons, this legislation enshrines that simple principle into law."[26] Even if this legislation doesn't pass, Senator Markey and Representative Lieu's efforts are alerting the public to the dangers of sole authority.

Despite the lack of confidence in President Trump that these efforts reflect, at least among Democrats, some had drawn unwarranted reassurance from the thought that there were "adults" in the administration, such as Secretary of Defense James Mattis, who would temper Trump's temperament. Senator Mark R. Warner (D-VA) called Mattis "an island of stability amid the chaos of the Trump administration."[27] Then senator Jeff Flake (R-AZ) told the *Washington Post* that "having Mattis there gave all of us a great deal more comfort than we have now."[28]

Mattis, who resigned from the Pentagon in late December 2018, reportedly told Strategic Command staff to tell him of any event that might lead to a nuclear alert being sent to the president, and "not to put on a pot of coffee without letting him know."[29] But just as Schlesinger had no formal control over Nixon, Mattis had none over Trump, even when he was the secretary of defense.

As concerns about Trump's nuclear authority grew, proposals to prohibit the first use of the bomb started to attract wider attention. Prohibiting first use (and leaving the option open for retaliation) would not end sole authority completely, but it would significantly limit the president's power to start a nuclear war.

In a November 2018 speech at American University, Senator Elizabeth Warren (D-MA) called for a "no first use" nuclear weapons policy. "To reduce the chances of a miscalculation or an accident, and to maintain our moral and diplomatic leadership in the world, we must be clear that deterrence is the sole purpose of our arsenal," Warren said.[30]

Soon thereafter, Senator Warren and Representative Adam Smith (D-WA), chairman of the House Armed Services Committee, proposed a bill that would establish in law that the United States would not use nuclear weapons first. "Our current nuclear strategy is not just outdated—it is dangerous," they said in a joint statement. The lawmakers said their bill would codify what most Americans already believe: that the United States should never initiate a nuclear war.[31]

The country is waking up to the frightening reality that when it comes to nuclear weapons, presidents have almost complete autonomy with essentially no institutional checks and balances. This authority was neither enshrined in the Constitution nor legislated by Congress.[32] It was not given but *taken* for themselves by the first presidents of the nuclear age. Today, the authority to use the bomb is a key part of presidential power and prestige. It has become an executive privilege like no other. And like any privilege, it can be taken away.

## HOW WE GOT HERE

The original rationale for presidential control over the use of nuclear weapons had nothing to do with the need to use them quickly, and everything to do with never using them again.

Back in 1945, at the dawn of the nuclear age, the United States had a monopoly on the bomb. No one else had one. After the US nuclear attacks on Japan, there was no risk that Japan would have a nuclear response. And for the next decade, until the first Soviet nuclear test in 1949 and long-range missile launch in 1957, there was no risk that the United States would be hit with a nuclear attack before it could respond. As a result, the driving concern for President Truman was not to ensure that the weapons could be launched quickly. Truman's priority was to guard against overzealous generals.

The conventional wisdom is that President Truman was in complete control of the bombings of Hiroshima and Nagasaki. In fact, Truman did not know when the bombs would be used (it depended on the weather) and may not have known that a second bomb would be dropped at all. According to historian Alex Wellerstein, "Truman was a president out of the loop, shouldering political burdens imposed by a military operating with its own priorities and agenda, with little civilian oversight into their day-to-day operations."[33]

After the atomic bombings, however, Truman realized that this was too important to delegate to the military. When Gen. Leslie Groves wrote to Gen. George Marshall that a third bomb could be ready for use in about a week, Gen. Marshall quickly replied that the bomb "is not to be released over Japan without express authority from the President."[34] The third bomb was never used, and presidential sole authority was born. From then on, nuclear weapons would be known as "the president's weapons."

Truman told Secretary of Defense James Forrestal that he did not want "to have some dashing lieutenant colonel decide when would be the proper time to drop one."[35] Henry Wallace, secretary of commerce, wrote in his diary that Truman thought the idea of killing "another 100,000 people was too horrible."[36]

Congress soon asserted civilian control of atomic weapons by passing the Atomic Energy Act in 1946. In Truman's words, "since a free society places the civil authority above the military power, the control of atomic energy properly belongs in civilian hands."[37] Unlike all other weapons of war, the Act gave control over nuclear weapons research and production to the president and civilian bureaucracy—and not to the military. But it did not discuss authority for using the bomb.

By late 1945, just months after using the bomb, Truman was seeking international control of atomic energy: "I have sought to eliminate atomic weapons as instruments of war, by seeking through the United Nations to put the control of the dangerous aspects of atomic energy beyond the reach of any individual nation." The world today would be very different if Truman had succeeded in this effort, but he did not. Truman's proposal was undermined by growing United States–Soviet distrust and was ultimately blocked by the Soviet Union's plan to make a bomb of its own (see chapter 9).

In July 1948, in a meeting with his advisors, Truman made his rationale for keeping the bomb to himself quite clear:

I don't think we ought to use this thing unless we absolutely have to.
It is a terrible thing to order the use of something that is so terribly

destructive, destructive beyond anything we have ever had. You have got to understand that this isn't a military weapon. It is used to wipe out women and children and unarmed people, and not for military uses. So we have got to treat it differently from rifles and cannons and ordinary things like that.[38]

The United States did not have a formal policy on who controlled nuclear use until September 1948, when the National Security Council declared, "The decision as to the employment of atomic weapons in the event of war is to be made by the Chief Executive when he considers such decision to be required."[39]

During the 1948–49 Berlin Blockade—one of the Cold War's first major crisis points—President Truman refused to shift control of the nuclear stockpile, then about fifty weapons, from the civilian Atomic Energy Commission to the Pentagon. He also refused to delegate his authority to use the weapons. At the same time, Truman declared that he would defend Berlin and "this Government is prepared to use any means that may be necessary."[40]

During the Korean War (1950–53), there was a public power struggle between Truman and Gen. Douglas MacArthur over control of nuclear weapons. To clarify things, the White House released a statement saying that "only the President can authorize the use of the atom bomb, and no such authorization has been given." This policy was carried over into the Eisenhower administration.

Today this authority is widely recognized. A 2013 Pentagon report states that "consistent with decades-long practice, the President, as Commander in Chief of the US Armed Forces, has the sole authority to order the employment of US nuclear forces."[41]

Sole authority is now generally accepted, but that does not mean it was inevitable or that it makes sense. As President Kennedy wrote in a 1962 classified memo, "From the point of view of logic there was no reason why the President of the United States should have the decision on whether to use nuclear weapons. History had given him this power."[42]

## EROSION OF CIVILIAN CONTROL

In the early years, the line between civilian and military control was clear and bright, right down to the weapons themselves. Back then, a bomb's nuclear components (plutonium and highly enriched uranium) were kept separate from its other nonnuclear parts. The Atomic Energy Commission (now absorbed into the Department of Energy), a civilian agency created by the Atomic Energy Act, controlled the nuclear parts, and the military got the rest. Only the president had the authority to transfer the nuclear parts to the military and then order their use—a two-step process.

But over time, the bright line began to blur. As atomic bombs grew in number and complexity, it became impractical to manually insert nuclear parts into bombs just before use. The introduction of ballistic missiles in the 1950s, first in ground-based silos and then on submarines, and the advent of thermonuclear warheads made keeping the nuclear components apart impossible. Nuclear warheads now had to be lighter and smaller, and once they were loaded onto the missiles they could not be easily accessed from outside. No more tinker toy bombs.

And once the Soviets had nuclear warheads on long-range missiles by the 1960s, everything began to change. The United States now had to worry, at least in theory, about "preemption." What if an atomic bomb exploded on Washington, DC, and killed the president and all those in the administration who might be next in line? Such a "decapitating strike" could theoretically make a US retaliation impossible.

After the Cold War, neither President Clinton nor Russian President Boris Yeltsin believed the other would launch a nuclear strike. Our two countries were working together, almost as allies: We cooperated in the dismantlement of eight thousand nuclear weapons, half in the former Soviet Union. I (Bill), as Clinton's defense secretary, personally oversaw the dismantlement of more than two thousand nuclear weapons in Ukraine and brought a Russian defense minister to Whiteman Air Base, where we destroyed a Titan ICBM silo. But, as we will see, this cooperative relationship would not last.

Eventually the bureaucratic struggle for nuclear control settled into the compromise we have today: the military requests the nuclear weapons from Congress, which authorizes civilians at the Department of Energy (more specifically, the National Nuclear Security Administration, NNSA) to build the warheads. The military then takes possession of the weapons, deploys them, and the president has sole authority to order their use.

## THE FOOTBALL

Truth be told, there is no "nuclear button." It is a metaphor that helps us understand the ease and speed of a presidential order to launch. Indeed, actually having a button at the ready for the president to press would raise all kinds of troubling possibilities, such as pressing it by mistake. Or spilling a soda on it. The closest thing we have to a nuclear button is actually a briefcase known as the "football," formally called the president's emergency response satchel.

President Kennedy had concerns about what would happen if he were away from the White House when he received information that might lead him to launch an immediate nuclear strike. Would emergency communications systems allow him to issue a launch order from anywhere in the world? If he called the War Room in the Pentagon to order such a strike, what would he say? And how would the duty officer verify the authenticity of the president? In response to Kennedy's questions, the Joint Chiefs of Staff designed the nuclear football.[43] In 1962, Kennedy helped formalize presidential sole authority when he decreed that the football would follow him at all times.[44]

The football, or satchel, is carried by a military assistant who is available to the president 24-7-365, and it contains a menu of nuclear strike options, a card with authentication codes, and a secure phone. These codes allow the president to issue orders to the National Military Command Center (NMCC) in the Pentagon. The codes are printed on a special card, called the "biscuit," that is supposed to be in the president's possession. (Presidents Carter and Clinton both lost their cards on two occasions.)

The military aides to the president, representing each of the services, alternate shifts by his side and sleep nearby, such as in the basement of the White House. They must always be within minutes of their boss, including riding in the same elevator. These tireless aides are often seen in the background of photos of the president, carrying the always present football (see pages 145–148).

If the president is dead, incapacitated, or otherwise unavailable, launch authority devolves first to the vice president. This process was put to the test on November 22, 1963. Army warrant officer Ira Gearhart, carrying the emergency satchel, was in the back of the president's motorcade in Dallas when Kennedy was shot. The Pentagon was alarmed at first due to uncertainty about Gearhart's whereabouts, but he was on Air Force One when Lyndon Johnson took the oath of office.[45]

During the attempted assassination of President Reagan in March 1981, the military aide carrying the football was separated from the president and did not accompany him to George Washington University Hospital. In preparation for surgery, Reagan's street clothes were removed and the biscuit was later found abandoned in a hospital plastic bag.[46]

If the vice president is also unreachable, launch authority moves down the chain of succession to the Speaker of the House, the President Pro Tempore of the Senate, the Secretary of State, the Secretary of the Treasury, and the Secretary of Defense. These are followed by the remaining members of the cabinet, in the order in which their departments were created.[47] It is notable that the Defense Secretary is sixth in line.

If you have ever watched a State of the Union address or presidential inauguration, you may have noticed that one member of the president's cabinet is always missing. This "designated survivor" stays at a separate location, presumably watching the speech on TV, in case there is an attack that incapacitates everyone else. This official is accompanied by a military aide with a nuclear football. During President Clinton's second inauguration, I (Bill) was the designated survivor—my family attended the inauguration along with the rest of the president's cabinet on the Washington Mall while I stayed at an undisclosed location. Fortunately, I never had to exercise any nuclear emergency protocols.

Since the end of the Cold War, both US and Russian leaders have questioned the need for the president to be followed by the football at all times. In October 1991, just after the failed coup against Soviet president Gorbachev, during which he lost control of his football (known in Russian as the *chemodanchik*, literally "small suitcase"), President George H. W. Bush wrote a note to his national security advisor, Brent Scowcroft. He asked, referring to the emergency satchel, "Does Mil Aide need to carry that black case now every little place I go?"[48] As Bush recalled, "With the Cold War over, I did not think it was necessary for the 'football' to go everywhere with me."[49] Scowcroft and others convinced him that it was still necessary.

According to recently declassified documents, President Clinton and Russian president Boris Yeltsin discussed their footballs on two occasions in 1994 and 1997. Surprising Clinton, Yeltsin suggested getting "rid of the nuclear footballs" so that aides no longer had to "drag" them around. Yeltsin saw the US football and the Russian equivalent as obsolete, saying, "If I need to communicate with the nuclear forces, I can pick up any number of telephone lines to do so."[50]

At first, Clinton responded jokingly, "What would my military aide do?" Seeing Yeltsin was serious, Clinton said, "When I took this job, I understood the symbolic importance that the football has in terms of civilian control of the military's decisions. It has nothing to do with you. It is a double-check that only a civilian, elected leader can make this decision."

Clinton's response is fascinating. He was saying that the football was not about a nuclear attack against Russia, but merely a symbol of civilian control. Not a threat, but a reassurance. This would appear to reflect Clinton's view that nuclear war with Russia was a remote possibility.

Yeltsin brought up his proposal at a second meeting in 1997 in Helsinki, after describing how, during heart surgery, he transferred the Russian football to Prime Minister Viktor Chernomyrdin. Yeltsin asked, "What if we were to give up having to have our finger next to the button all the time? We have plenty of other ways of keeping in touch with each other. They always know where to find us, so perhaps we could agree that it is not necessary for us to carry the *chemodanchik*."

Clinton replied, "Well, I'll have to think about this. All we carry, of course, are the codes and the secure phone." Deputy Secretary of State Strobe Talbott said that it was better for presidents "to have these devices with you at all times rather than to have the function assigned to a computer somewhere or to anyone else."

It is not clear, however, that Yeltsin was suggesting that these functions be transferred to others. Instead, he might have been proposing that the option for quick launch be discontinued. This could have been an important opportunity to back away from the nuclear brink. But Clinton was not receptive.

## THE OPTIONS

The president's options for nuclear strikes are elaborated at length in a "black book" contained in the football. These are also summarized in a one-page cartoon-like menu, thanks to President Jimmy Carter, who found the long version too complicated to comprehend within the few minutes allowed.

What are the nuclear strike options available to the president? At any time, US nuclear forces can deliver hundreds of warheads to targets around the globe. With a few days of preparation, that number would grow to more than one thousand. In either case, the US arsenal could decimate an adversary's nuclear forces, war-supporting industries, and key command posts with top political and military leadership.[51]

Once the president issues orders, the well-oiled military machine takes over. The orders travel down the chain of command until they reach launch officers in missile silos and on submarines, who have trained for this moment. Only a coordinated, widespread mutiny could stop this process from reaching its grim end. The whole process would take just minutes.

The president can order the use of nuclear weapons without the input or consent of senior advisors, such as the secretaries of Defense or State, the chairman of the Joint Chiefs of Staff, or the vice president. Unless the president has given prior approval for advisors to be notified, none would need to be involved with the decision.[52]

## THE NEED FOR SPEED

It is now firmly established that the US civilian leadership controls the use of nuclear weapons, and the military does not. This is a good thing. But this begs another question. Why just the president? Why not involve other civilian leaders, like members of Congress?

One answer is the cold, hard logic of a nuclear attack. A Russian nuclear missile strike would reach the United States in thirty minutes or less, and there may simply be no time to consult with Congress, a subset of Congress, or even the president's closest advisors. The president can be encouraged to consult as much as possible, but given current policy it would seem impractical to require consultation.

We have lived under this theoretical threat of immediate destruction—an atomic Pearl Harbor—since the 1960s. But, as we shall see, it is not necessary for deterrence that a US response to a nuclear attack be immediate. And if we want to avoid blundering into nuclear war, the more decision time we can give a president, the better.

In addition to presidential sole authority, the perceived need to launch US nuclear forces before a Russian attack arrives drives a host of other dangerous policies. Primary among them is the US policy to be able and ready to launch ICBMs before they can be destroyed in their silos. The fixed locations of US ICBMs are well known, and today, as a result of improvements in the accuracy of Russian ICBMs, the US ICBMs are highly vulnerable. By comparison, US submarines at sea would not be destroyed by a Russian strike (since their locations are secret), nor would US strategic bombers if they were sent aloft at the first sign of a crisis. Submarines at sea can cruise for weeks or months after an attack, and bombers can stay airborne for hours and be recalled before they launch their weapons.

For US ICBMs to survive a first strike from Russia, they must be launched before the attack arrives. Thus, they must be on alert and ready for launch 24-7. Not only that, but there must also be a warning system in place that can detect the Russian launch with high confidence so that the US ICBMs can be launched on the basis of that warning.

This is where the trouble begins. The president has four ways to launch nuclear weapons: an unprovoked first strike (no international crisis), a preemptive first strike (during a crisis when it may be feared the other side is about to strike), a launch on warning of attack (missiles are reportedly headed our way), or a retaliatory attack (after an attack has landed). To preempt a possible nuclear attack, there must be highly reliable intelligence that the adversary is about to launch—but has not yet done so. To launch ICBMs on warning of attack means that the attack, if real, has not yet landed. Therefore, if US ICBMs are launched and the reported attack turns out to be a false alarm, the United States would just have started a nuclear war by mistake (as described in the opening scenario). The weapons cannot be recalled, and it is unlikely that Russian leaders could be convinced not to retaliate on the basis that it was all a big mix-up. In other words, the end of civilization would be at hand.

The United States developed the ability to launch on warning in the early 1970s, under the Nixon administration.[53] President Carter's Presidential Directive 59 stated: "While it will remain our policy not to rely on launching nuclear weapons on warning that an attack has begun, appropriate pre-planning, especially for ICBMs that are vulnerable to a preemptive attack, will be undertaken to provide the President an option of so launching." Similarly, under President Reagan, the policy was that "the United States does not rely on its capability for launch on warning or launch under attack to ensure credibility of our deterrent. At the same time, our ability to carry out such options complicates Soviet assessments of war outcomes and enhances deterrence."[54]

This wonky policy prose hides a deeply ingrained preference among nuclear war planners for launching US nuclear weapons first—either preemptively or on warning of attack—rather than after an attack. As far back as Gen. Curtis LeMay and the Strategic Air Command, nuclear warfighters intended to gain warning of an attack and "beat him to the draw" by launching a preemptive strike.[55]

From a war planner's point of view, a first strike is a straightforward affair: carry out the preprogrammed war plan. But retaliation after an attack is a planner's nightmare. All or most ICBMs would be destroyed. Communications would be at best unreliable, and coordination of forces next to impossible. Since no one has ever been able to practice communications during a nuclear war, thankfully, no one really knows what would happen.

"However demanding our no-notice exercises, they could never realistically replicate the equally problematic circumstances of an actual nuclear attack," said former STRATCOM commander Gen. George Lee Butler, "during which men and machines would be tested under the most demanding and terrifying conditions imaginable."[56]

But from a president's point of view, launch on warning raises the possibility of starting nuclear war—by mistake. President Kennedy's defense secretary Robert McNamara called launch on warning "insane," arguing that "there's no military requirement for it."[57]

What makes this situation so dangerous is that we know that the US early warning system is fallible. The United States has had at least three false alarms, and Russia at least two. We also know that presidents are fallible. Humans are prone to mistakes, and their machines can fail. (More on this last point in chapter 3.)

It is no wonder that all presidents in the nuclear age have recoiled at the thought of using the bomb a third time. The more powerful the weapons have become, the less useful as weapons they are.

"As the weapons began to rain down, the final truth would be laid bare," Gen. Butler wrote about what a real nuclear war would be like. "There would be no winners in this merciless exchange; once deterrence failed there would be nothing left to fight for, only retribution to be exacted. Now the only certainty was that, in the face of utter desolation, the living would surely envy the dead."[58]

It is interesting to note that two recent US presidents, George W. Bush and Barack Obama, both talked during their campaigns about the risks of false alarms and the need to take nuclear weapons off high alert. Bush declared that "the United States should remove as many weapons as possible from high-alert, hair-trigger status," and argued that the capability for a "quick launch within minutes of warning" was an "unnecessary vestige of cold-war confrontation." Bush added that "keeping so many weapons on high alert may create unacceptable risks of accidental or unauthorized launch."[59] Obama promised to take Minuteman missiles off alert, warning that policies like launch on warning "increase the risk of catastrophic accidents or miscalculation."

Unfortunately, when these leaders had the authority and opportunity to significantly change these policies, they did not. US Minuteman missiles, armed with nuclear warheads, sit in their silos today, ready to launch in minutes, at the direction of the commander in chief.

The outdated Cold War policies discussed here—sole authority, first use, and launch on warning—are highly dangerous and undermine the security of the United States. We would be safer without them. These misguided policies are all built on one assumption: that Russia is planning to launch a disarming first strike. As we explain in the next chapter, there is no basis for this assumption. And without this foundation, the core concepts of US nuclear policy fall away.

# CHAPTER 2

# BOLT FROM THE BLUE

*There is nothing in the world that the Communists*
*want badly enough to risk losing the Kremlin.*
—President Dwight Eisenhower[1]

I t's October 14, 1962. The Kennedy administration just discovered that the Soviet Union had secretly moved nuclear-capable missiles into Cuba that could reach the US East Coast. At the time, the United States had about five thousand nuclear warheads; the Soviets had three hundred. But even with this seventeen-to-one numerical superiority, the Kennedy administration did not believe it had the capability to launch a successful first strike. There was still an unacceptable risk that Russia could launch a counterstrike. Just a few of its massive warheads would inflict substantial, devastating damage to the United States. New York, Washington, Chicago, Los Angeles, gone in a flash of light. So even in this highly lopsided situation, Washington was "deterred" from attacking Moscow.[2]

Soon thereafter, Russia built up its forces and closed the gap, making it even less likely that the United States would consider a first strike. As for Moscow, it never had nuclear superiority of this magnitude over the United States. Therefore, could Russia ever have been confident in its ability to launch a nuclear surprise attack against Washington and survive? We think not.

According to a declassified 1995 Pentagon report, Soviet military leaders "understood the devastating consequences of nuclear war" and believed that the use of nuclear weapons had to be avoided at "all costs." A 1968 Soviet Defense Ministry study showed that Moscow could not win a nuclear war, even if it launched a first strike. And in 1981, the Soviet General Staff concluded that "nuclear use would be catastrophic."[3]

Today, climate science tells us that a full-scale nuclear attack by either side would be suicidal *even if there were no retaliation*. The impact on the global climate from a unilateral attack would be so extreme as to eventually doom the attacker.[4]

Yet the US president has sole authority to launch nuclear weapons primarily to allow him or her to respond quickly—within minutes—to a surprise attack, a "bolt from the blue" from Russia. But what if Russia has no such intentions? What if leaders in Moscow have no interest in starting a nuclear war that they cannot possibly win? In other words, what if they are rational, like us?

For decades, the United States has been prepared for a surprise Russian nuclear attack that never arrived and, in all likelihood, never will.

---

During the Cold War, many believed that the Soviet Union was planning a disarming surprise nuclear attack on the United States. Thus, the US nuclear arsenal was sized to be able to withstand such an attack and still retaliate with overwhelming force. Missiles were deployed on high alert, ready to launch, to ensure that a surprise attack could not succeed in disarming the United States. This strategy was expensive, requiring large numbers of deployed weapons. But US defense officials believed that this strategy was necessary to deter the surprise attack the Soviets were presumably planning. At the same time, Moscow made the same assumptions about Washington.

Looking back at the Cold War, we find no compelling evidence that either side would have launched a surprise attack. As President Eisenhower concluded as early as 1955 after meeting with Soviet leaders in Geneva, "There seems to be a growing realization by all that nuclear warfare, pursued to the ultimate, could be practically race suicide."[5] And as former CIA director Bob Gates wrote in 1996, "In fact, very few in Washington thought there was even a remote chance that the Soviets would suicidally throw the dice that way."[6] Yet even those who were skeptical about US assumptions on Soviet strategy believed that the costs required to deploy huge deterrent forces provided "good insurance." And that was probably valid reasoning if the insurance costs were measured only in dollars. But US nuclear forces and policies have costs beyond that. These terrible Cold War weapons create a *danger of their own*: they increase the likelihood that these forces could be used by accident, through a political or technical miscalculation.

Gen. George Lee Butler wrote in 2016 that "the presence of these weapons inspired the United States and the Soviet Union to take risks that brought the world to the brink of nuclear holocaust. It is increasingly evident that senior leaders on both sides consistently misread each other's intentions, motivations, and activities, and their successors still do so today. In my own view . . . nuclear deterrence in the Cold War was a dialogue of the blind with the deaf. It was largely a bargain we in the West made with ourselves."[7]

The fear of a surprise nuclear attack drove US and Soviet nuclear policy throughout the Cold War. Neither nation believed that it would ever itself initiate such an attack, yet at times believed that the other side was planning to. We strongly believe that the risk of a surprise attack was and is significantly smaller than the risk of stumbling into a nuclear war.

If we are right, then throughout the Cold War and still today the two superpowers have been focused on the wrong threat. We have been undermining our own security. In the words of Mikhail Gorbachev, "Everything we had been doing was an error."[8]

## A NUCLEAR PEARL HARBOR?

Even before the atomic bomb, both the United States and the Soviet Union had bad memories of surprise attacks. On June 22, 1941, Hitler's Germany invaded the Soviet Union in Operation Barbarossa. Later that year, the United States suffered Japan's attack on Pearl Harbor on December 7. These attacks left deep impressions on the national psyches and reinforced an entrenched fear of a "bolt from the blue."

The advent of nuclear weapons made the prospect of a surprise attack all the more terrifying. Once nuclear warheads were placed atop ballistic missiles in the 1960s, both nations had the technical ability to turn the other into smoldering ruins in thirty minutes or less. Both nations spent trillions of dollars making sure that enough nuclear weapons, along with the decision makers and military officers to use them, would survive a surprise attack to allow a forceful and deadly retaliation.

The ability to retaliate was and is the only defense against the bomb (our missile defenses are not capable of defending against a large-scale attack, as we discuss in chapter 8). We feared that if the Soviet Union believed that it had the capability to attack the United States and destroy all of its nuclear weapons, then it would be tempted to do so. At the same time, if there is no reliable way to prevent nuclear retaliation after a first strike, then there is no incentive to launch one. And in nuclear war, even a relatively small retaliation could mean many tens of millions dead. It would be national suicide, and could have consequences far beyond our borders.

There were times early in the Cold War when military planners from both sides considered first strikes. In the late 1940s, before the Soviets had the bomb, there were calls for the United States to attack Russia before it had the means to retaliate. In 1948, the Joint Chiefs of Staff approved the first war plan against the Soviet Union that would have launched an "atomic blitz," dropping more than a hundred bombs on seventy Soviet cities, if the United States was losing a conventional battle. President Truman did not like the plan and told the Joint Chiefs to prepare a new one without the bomb. But once the Soviet blockade of Berlin began soon after, nuclear weapons became part of a possible American response to a Soviet invasion of western Europe.[9]

In the 1950s, once the Soviets had the bomb, the strategic balance was unstable, and each side had an incentive to strike first if an attack seemed imminent. For example, Gen. Curtis LeMay, the head of Strategic Air Command, said in 1954 that "I believe that if the US is pushed in the corner far enough, we would not hesitate to strike first." In 1961, Gen. Thomas Power argued that "if a general atomic war is inevitable, the US should strike first."[10] And, in 1962, the US Air Force sought to achieve a first-strike capability, to initiate and win a nuclear war, which Secretary of Defense McNamara fought against and won.[11]

After 1962, the Soviets had no hope of prevailing with a disarming first strike, as the United States by then had built hardened ICBM silos that would have taken thousands of Soviet ICBMs to destroy. Even more important, the United States began deploying submarine-launched Polaris missiles that were invulnerable to an attack from Soviet ICBMs.

By the time of the Cuban Missile Crisis, both sides knew that nuclear war was unwinnable and would be a disaster for all concerned. Even though the United States had a clear numerical advantage in strategic forces, it did not matter. "What we knew about Soviet nuclear forces at the time was simply that they were large enough to make any nuclear exchange an obvious catastrophe for Americans too," wrote Kennedy's national security advisor McGeorge Bundy. "We had to assume that in any nuclear exchange, no matter who started it, some of these missiles and bombers would get through with multimegaton bombs. Even one would be a disaster. The fact that our own strategic forces were very much larger gave us no comfort."[12]

As McNamara said in late 1962, "I am convinced that we would not be able to achieve tactical surprise, especially in the kinds of crisis circumstances in which a first-strike capability might be relevant. Thus, the Soviets would be able to launch some of their retaliatory forces before we had destroyed their bases."[13]

Nuclear deterrence was born, better known as Mutually Assured Destruction, or MAD.

A new war plan, the Single Integrated Operational Plan for Fiscal Year 1962, or SIOP-62, came into effect on April 15, 1961. Joint Chiefs of Staff Chairman Gen. Lyman L. Lemnitzer explained to President Kennedy that the execution of SIOP-62 "should permit the United States to prevail in the event of general nuclear war." Yet Gen. Lemnitzer also sounded a strong cautionary note, informing the president that "under any circumstances—even a preemptive attack by the United States—it would be expected that some portion of the Soviet long-range nuclear force would strike the United States."[14] That is, even as we "prevailed," many tens of millions of Americans would die.

By 1963, McNamara had concluded that "the increasing numbers of survivable missiles in the hands of both the United States and the Soviet Union are a fact of life. Neither side today possesses a force which can save its country from severe damage in a nuclear exchange. Neither side can realistically expect to achieve such a force for the foreseeable future."[15]

Some cold warriors have held on to the notion that, even if one accepts that Russia will not launch a first strike in peacetime, it might still launch one in a crisis. Moreover, some argue that the United States should be prepared to launch first if US leaders perceive that Russia is about to attack. Some see a preemptive, "damage-limiting" strike as the moral thing to do, as it could reduce the damage the United States might suffer from a Russian attack. But of course, our perceptions that Russia was about to attack could be in error, in which case the presumed "morality" of a first strike would have become an obscene immorality.

When you realize the scale of death involved in even a "damage-limiting" strike, the sheen of sanity wears off such dangerous ideas. As Henry Kissinger said in 1976, "Our estimates are that we would lose as many as 125 million people in a nuclear war if we didn't strike first, and 110 million if we did."[16] Given these awful choices, the obvious best path is to never push a crisis to the brink of nuclear war.

In 1971, Joint Chiefs Chairman Adm. Thomas Moorer explained that as long as the United States had survivable and redundant nuclear forces that could cause enormous destruction, "the other fellow" would not be tempted "'to give it a try' to see if he can get away with it." Moorer said it was "highly unlikely that the United States would ever preempt and make a first strike." A first strike, he added, would happen only if Washington had "positive assurance" of an imminent attack but "I don't see how we could have such assurance."[17]

As Gen. LeMay's successor as SAC commander, Gen. Thomas Power noted that after a US first strike, "some of their bombers and missiles would escape destruction and succeed in mounting a counterattack, exacting a high price with their nuclear payloads." Power believed that a US preemptive strike was unlikely because "so long as there is the slightest hope that we can prevent a Soviet attack through diplomatic means or a strong posture of deterrence, our government, backed by the majority of the American people, would in my opinion be opposed to more drastic measures."[18]

Gen. Richard Ellis, a successor to LeMay and Power, had this to say in 1979 about the effects of all-out nuclear war: "It would be a catastrophic event of such magnitude that I don't think the human mind could understand it . . . it would be difficult to distinguish . . . which would be the victor or the vanquished."[19]

## "SANE LEADERS DO NOT PLAY RUSSIAN ROULETTE"

It was clear early on to policy makers in the United States that neither side would win a nuclear war and that such a war should never be waged. Although US presidents might issue thinly veiled nuclear threats, they had no appetite to actually use nuclear weapons except in retaliation. The key thing was to maintain a nuclear force large enough such that Moscow would be deterred from launching an attack. As McNamara put it in 1968: "If the United States is to deter a nuclear attack on itself or its allies, it must possess an actual and a credible assured-destruction capability." And to McNamara, such a capability meant "the certainty of suicide to the aggressor, not merely to his military forces, but to his society as a whole."[20]

Some might quibble with McNamara's requirement of "certainty of suicide." Bundy believed that a moderate *risk* of suicide was an adequate deterrent. We agree. As Bundy wrote, "Sane leaders do not play Russian roulette."[21]

But not everyone was satisfied with the goal of second-strike retaliation. And even if they were, it begged the question of how much was enough? How many nuclear weapons, and of what kind, were required to deter Soviet aggression?

Cold War tensions and a lack of transparency into Soviet decision making made it difficult to get an accurate estimate of Soviet nuclear forces and what Moscow planned to do with them. During the Cold War, US intelligence services made their best guesses, and were often wrong.

If you cannot accurately estimate what the other side has, then the "prudent" thing to do is to assume the worst case—the other side has more than you. So, you build up to catch up. The other side sees this buildup and does the same. This arms race based on bad information and worst-case assessments led the United States and the Soviet Union to build well over 70,000 nuclear weapons since 1945, a staggering and unjustifiable amount.[22] Today, thanks to bilateral and unilateral reductions, both nations together retain fewer than 13,000 warheads, still more than enough to destroy the entire planet.

## THE SO-CALLED MISSILE GAP

A good example of this misguided arms race dynamic was the so-called missile gap. The Soviets launched the first satellite into orbit on October 4, 1957. This showed the existence of a long-range ballistic missile powerful enough to boost the satellite into space. This event was a shock to the US political system and to prevailing assumptions about American technical advantage. The United States had underestimated the technical capabilities of the Soviets and had considered itself so far advanced that it had no serious rival. No longer.

After the launch, the *Washington Post* breathlessly declared that only through an all-out effort could the United States "hope to close the current missile gap and to counter the worldwide Communist offensive in many fields and in many lands."[23] In 1959, Senator Stuart Symington declared that "in three years the Russians will prove to us that they have 3,000 ICBMs."[24] Soviet premier Nikita Khrushchev said he was building missiles "like sausages" and that they were in "mass production." John F. Kennedy used the missile gap as an issue in his 1960 presidential campaign against Richard Nixon, saying, "Our Nation could have afforded, and can afford now, the steps necessary to close the missile gap."[25]

In 1961, a young Henry Kissinger wrote, "The missile gap in the period

1961–1965 is now unavoidable." He continued, "The vulnerability of our retaliatory force will create major opportunities for Soviet nuclear blackmail—even to the extent of threatening direct attacks on the United States."[26] The exaggeration of nuclear threats to serve political agendas has not gone out of fashion; the George W. Bush administration used it to great effect to justify the US invasion of Iraq in 2003. The war was a strategic disaster for the United States and the Middle East, and no nuclear weapons were ever found.

President Eisenhower did not buy Kissinger's alarmist prediction. According to Bundy, Eisenhower "never believed for a moment that the Soviet government would deliberately choose a nuclear war, and so he saw no need for any American president to be moved by Soviet nuclear blackmail."[27]

In Eisenhower's words, "There is nothing in the world that the Communists want badly enough to risk losing the Kremlin."

As it turned out, Khrushchev was bluffing. The missile gap was a myth.

In August 1959, CIA director Allen Dulles convened a special panel and invited me (Bill) to join. Named after its chair, Pat Hyland, president of Hughes Aircraft, the Hyland panel reviewed all of the available intelligence, and got briefings from the CIA, the NSA, and all three military services. We concluded unanimously that the Soviet ICBM program had only a few deployed missiles and was not rapidly expanding. There *was* a missile gap, but it was in *our* favor. However, this report was highly classified and not made public until decades later and could not be used to dampen the public concern at the time.

At about the same time the Hyland panel was meeting, a new and remarkable intelligence capability was being deployed by the United States: the Corona satellites. These satellites were in orbit over the Soviet Union every day taking sweeping panoramic pictures; the United States penetrated the Iron Curtain by flying over it. As this program became fully operational, the number of ICBMs deployed in the Soviet Union would become known to American intelligence with good accuracy.

Soon after the election, in February 1961, Secretary of Defense Robert McNamara stated that there was no evidence of a large-scale Soviet effort to build ICBMs. Satellite overflights had found only a relatively few operational launch sites, and by September, a National Intelligence Estimate concluded that the Soviets had no more than twenty-five long-range ballistic missiles and would not possess more anytime soon. At the time, the United States had more than twice that number.

During McNamara's first press conference after being sworn in as secretary of defense, he was asked about the missile gap. McNamara replied, "Oh, I've learned there isn't any, or if there is, it's in our favor."[28] Even so, there was widespread concern that the Soviet Union was preparing to win a nuclear war. "In fact, until the mid-1960s, writings of Soviet military officials consistently maintained that the only conflict possible between the great powers was an all-out nuclear war," McNamara wrote. "They asserted, moreover, that it was possible to prevail in such a conflict, and they urged the military and social preparations necessary to ensure that the USSR emerged triumphant from any nuclear conflict."[29]

All the same, according to Bundy, even if there had been a missile gap it would not have led to an increased risk of surprise Soviet attack: "Survivable American forces would still present a wholly inescapable risk of wholly unacceptable retaliation, and the Soviet leaders would still be men who simply would not take such risks in the absence of immediate and catastrophic danger to themselves."[30]

The mythical missile gap reappeared in 1974, this time called the "window of vulnerability." It was promoted by University of Chicago professor Albert Wohlstetter, who accused the CIA of systematically underestimating Soviet ICBM deployments. In 1976, Paul Nitze, a former arms control negotiator, picked up this theme and warned that the Soviets were not seeking nuclear parity with Washington, but "will continue to pursue a nuclear superiority that is not merely quantitative but designed to produce a theoretical war-winning capability."[31]

At the time, claims that the Soviets had plans to start and win a nuclear war could not be proven or disproven. But to many insiders, even during the Cold War, they did not make sense. How could Soviet leaders, who were rational people, possibly believe that it would be in their interests to start a nuclear war that both sides would clearly lose? To help answer this question, in 1976 the CIA organized an extraordinary examination of Soviet intentions. It set up two teams to assess the same intelligence, one from inside the CIA (Team A) and the other made up of outside experts (Team B). The outsiders were led by Harvard professor Richard Pipes, a longtime critic of Russia. The Team B report concluded that Soviet leaders "think not in terms of nuclear stability, mutual assured destruction or strategic sufficiency, but of an effective nuclear war-fighting capability."[32] To Pipes and others like him, the Soviets were preparing to start and win a nuclear war.

The CIA did not agree, finding that the Soviets "cannot be certain about future US behavior or about their own future strategic capabilities relative to those of the US." In other words, it could not be clear to Moscow that it had any kind of strategic advantage. As the State Department's top intelligence official put it, Soviet leaders "do not entertain, as a practical objective in the foreseeable future, the achievement of what could reasonably be characterized as a 'war winning' or 'war survival' posture."[33] Thus, the Soviets could not be confident about winning or even surviving a nuclear war.

By 1977, Soviet premier Leonid Brezhnev admitted the possibility of a major war between East and West that did not use nuclear weapons. In 1982, the Soviets declared for the first time that they would not use nuclear weapons first. Speaking at the United Nations, Brezhnev said that the USSR "assumes an obligation not to be the first to use nuclear weapons." The Soviets were also asserting by this time that "there will be no victors in a nuclear war."

The debate raged on. Pipes wrote an article in 1977 called "Why the Soviet Union Thinks It Could Fight and Win a Nuclear War." Nitze, Pipes, and others helped establish an advocacy group called the Committee on the Present Danger (CPD) to raise public concern about the Soviet nuclear buildup and to defeat the second Strategic Arms Limitation Treaty (SALT II), then in negotiations under President Jimmy Carter. On the committee's board of directors sat former California governor Ronald Reagan.

During the Carter administration, I (Bill) served as undersecretary of defense for research and engineering. My task was to increase the capability of US conventional forces through technology; that is, to *offset* the quantitative superiority of the Red Army with American qualitative superiority. The three key elements of the offset strategy were smart weapons, smart intelligence systems, and stealth aircraft. This effort, which was strictly focused on our conventional forces, proceeded with the highest priority in the late 1970s, and was continued with high priority by the Reagan administration in the 1980s. It was remarkably successful, as demonstrated in the first Gulf War, when American military forces crushed a large Iraqi army in four days. So, the offset strategy, which was developed to defeat the Red Army, happily was never used against the Russians; it was instead used against the Iraqi army, which had been largely equipped and trained by the Red Army.

## NUCLEAR MODERNIZATION

In addition to major efforts to bolster US conventional forces, the Carter administration also pursued a modernization of nuclear forces. I (Bill) supported the upgrading of our submarine forces from Polaris to Trident, because I believed that our submarines were the key to an invulnerable nuclear force. And I launched a program for a major new stealth bomber, the B-2, which would be a significant addition to our conventional forces but could also carry nuclear bombs. But I only reluctantly continued the MX missile, which had been started in President Carter's administration in response to arguments that we were falling behind the Soviet Union.

The MX missile was a huge, ten-warhead ICBM to replace Minuteman. Because placing ten warheads on a silo-based missile would present an inviting target for preemptive attack, we tried to find a more survivable basing mode for the MX. Without a doubt, trying to resolve this issue, ultimately unsuccessfully, was the most frustrating experience I had as undersecretary.

The perceived need for parity did more to drive the arms race than the need for deterrence. We could achieve deterrence without parity. But it was the constant push for parity that drove both sides to keep building up and up. Similarly, the need for the "triad" of land-based, sea-based, and air-based weapons was and is more about politics than deterrence. We believe that the United States could have confidence in a deterrent force of only invulnerable submarine-based weapons, backed up with an insurance policy of bombers that could carry out either conventional or nuclear missions.

But in addition to arsenal size, the administration also had to make sure that the US arsenal could survive a first strike and be able to respond with a devastating counterstrike. It turned out to be infeasible to make the ICBMs survivable, but we built considerable survivability into the rest of the nuclear force.

Thus, I undertook major actions to modernize those parts of our nuclear arsenal that could be survivable, even though I did not believe that the Soviets were actually planning a disarming first strike. I worked on the Trident submarine program, developed the Air-Launched Cruise Missile for the B-52 bomber as an alternative to the B-1 bomber (which I canceled because it would not have been effective against Soviet air defenses), and started development of the B-2 stealth bomber.

The Committee on the Present Danger (CPD) raised the concern that a Soviet first strike could destroy US Minuteman III ICBMs in their silos, which I found greatly exaggerated. First, our silos would protect the ICBMs from anything other than a direct or near-direct hit. But based on our intelligence, we did not believe that Soviet ICBMs were accurate enough to give Russian leaders confidence that such an attack would destroy US missiles in their silos. Second, even if the Soviets were to develop such high accuracy, they could not rule out that we would launch our missiles before their ICBMs arrived, a practice called launch on warning. The US warning system was (and is) good enough to provide a ten- to fifteen-minute warning, and Minuteman missiles can be launched in a few minutes. And while I had great concerns for the dangers of launch on warning, concerns that were greatly amplified by my personal experiences with false alarms, I believed then that launch on warning was worth the risk. Now, with the Cold War long over, I believe that launch on warning is *not* worth the risk, given the ever-present possibility of a false alarm (see chapter 3).

We looked at basing the MX on airplanes, trains, and trucks, and underwater, all of which proved to be complex and expensive. Finally, we decided to build 200 MX missiles based in 4,600 silos in Nevada and Utah and planned to move the missiles around so the Soviets would not know where they were. Unsurprisingly, the citizens of Nevada and Utah rejected this plan, and I regret letting the prophets of doom stampede me into supporting this unwise project.

The Reagan administration was also unsuccessful with novel basing options and finally decided to deploy the MX in fixed silos. This did not undermine the overall US deterrent, which had survivable warheads on submarines and bombers. Although we can see bureaucratic politics at play here, what the Soviets saw was the US deploying ten-warhead missiles in vulnerable fixed silos, which to them made sense only as first-strike weapons. The MX's vulnerability (indeed, it was practically an invitation to a Soviet preemptive strike) did not matter if we intended to use the MX first. (The first Bush administration, after it had negotiated reductions with the Soviets, wisely retired the MX, even though it was the newest ICBM system we had.)

During the 1980 election campaign, Reagan criticized President Carter for seeking SALT II, which would have limited the number of offensive nuclear weapons on both sides. CPD was arguing that the treaty would make the United States more vulnerable to a Soviet surprise attack. This all came to a head in a national TV debate on SALT II. I was one of a team of three supporting ratification of SALT II. Paul Nitze led the team opposed to the treaty. It was an incredibly nasty debate, falling as it did in the middle of a presidential election that was still too close to call. Senator John Culver (D-IA, soon to be voted out of office in the Reagan landslide) opened the debate with a stirring defense of the treaty. Paul Nitze began his response by saying, "Senator Culver has given you some fine reasons for supporting this treaty. The only problem is, they were all lies!" The debate went downhill from there.

After the debate, President Carter asked me to meet with individual senators, one by one, to brief them on the treaty. I was about halfway done with those briefings, and confident of success, when Carter pulled the treaty from Senate consideration after the Soviets invaded Afghanistan in late 1979. The treaty was never ratified.

## "THE EVIL EMPIRE"

President Reagan took office in January 1981 and, by his actions, convinced the Soviets that he just might launch a nuclear first strike against Moscow. Reagan called Russia "the evil empire," increased US defense spending, and set out to build up US nuclear forces. In 1983, he announced plans to develop a nationwide defense against ballistic missiles called the Strategic Defense Initiative, or SDI, ridiculed by critics as "Star Wars." The Soviets saw this as part of a first-strike plan, where the United States would attack and use missile defenses to blunt a smaller Soviet retaliation.

Soviet premier Yuri Andropov said that Reagan was "inventing new plans on how to unleash a nuclear war in the best way, with the hope of winning it."[34] Reagan said, "Unlike us, the Soviet Union believes that nuclear war is possible. And they believe it's winnable . . ."[35]

The implausible theory behind this belief was as follows: The Soviets would launch a surprise attack, destroying most US ICBMs; surviving US submarines and bombers would be able to retaliate against Soviet cities, but US leaders would not do so due to fear of a Soviet counterattack against US cities; thus, the United States would have to yield to Soviet demands. By merely threatening such an attack, the Soviets could dictate US policy.

This, of course, was hogwash. "Those who accept the first-strike scenario view the Soviet ICBMs and the men who command them as objects in a universe decoupled from the real world," McNamara wrote in 1986. This theory also assumes the Soviets have complete confidence in their complex weapons systems, which have never been tested in wartime; that US satellites and other intelligence sources would miss all preparations for such an attack; that the United States would not launch its missiles once the attack was detected; and—most unbelievable of all—that the United States would not retaliate with hundreds of submarine- and bomber-based weapons, killing millions of Soviet citizens. This theory assumes Soviet leaders to be irrational.

"Only madmen would contemplate · such a gamble," McNamara wrote. "Whatever else they may be, the leaders of the Soviet Union are not madmen."[36]

As McGeorge Bundy put it: "Think tank analysts can set levels of acceptable damage well up in the tens of millions of lives. They can assume that the loss of a dozen great cities is somehow a real choice for sane men. They are in an unreal world. In the real world of political leaders—whether here or in the Soviet Union—a decision that would bring even one hydrogen bomb on one city of one's own country would be recognized in advance as a catastrophic blunder; ten bombs on ten cities would be a disaster beyond history; and a hundred bombs on a hundred cities are unthinkable."[37]

Meanwhile, Andropov was particularly concerned about US plans to deploy medium-range, nuclear-armed missiles in Europe that could reach Moscow in fifteen minutes and, he feared, decapitate Russian leadership and thereby make nuclear retaliation impossible. (This was in response to the Soviet deployment of SS-20 medium-range missiles aimed at European NATO members.) Soviet paranoia reached a peak in November 1983, just as NATO was launching a military exercise called Able Archer, designed to simulate nuclear war in Europe. The exercise may have been seen by Moscow as preparations for a real attack.

As for the "window of vulnerability," a 1983 report by the President's Commission on Strategic Forces led by Lt. Gen. Brent Scowcroft concluded that there was no such thing. I (Bill) was on this commission and found the discussions to be interesting and focused, thanks to the leadership of Scowcroft. The commission consisted of a cross-section of American security experts from both parties but had no difficulty arriving at a unanimous conclusion that the United States was not facing a window of vulnerability. The commission found that no combination of attacks from Soviet submarines and land-based ICBMs could destroy US bombers, missiles, and submarines. The commission did not examine whether there was any evidence that the Soviet Union was even trying to achieve such a capability, but all evidence available to us now suggests that they were not. So, this huge political uproar on the so-called window of vulnerability, with its attendant consequences on our politics and our force structure, was over a myth.

"Misperceptions such as the mythical missile gap and the window of vulnerability can be very costly indeed," concluded McNamara. "They can lead to inflated defense budgets, increased suspicions between East and West, and—in political or military crises—to misjudgments about the use of military force."[38]

Former CIA director Gates wrote that "by the early 1980s, [the Soviets] saw strategic weapons being deployed and new programs undertaken that they believed could provide the United States a first-strike capability." He concluded, based on intelligence sources, that Moscow took the threat of a US preemptive nuclear attack "very seriously" in 1983–1984. According to one of Gates's sources, "Few officials with direct experience of life in the West took the threat of a US first strike seriously, but in senior party circles such an eventuality was widely perceived."[39]

President Reagan later wrote in his memoirs that he was surprised that Soviet leaders would fear a US first strike, and that fear undermined US security:

> During my first years in Washington, I think many of us in the administration took it for granted that the Russians, like ourselves, considered it unthinkable that the United States would launch a first strike against them. But the more experience I had with the Soviet leaders and other heads of state who knew them, the more I began to realize that many Soviet officials feared us not only as adversaries but as potential aggressors who might hurl nuclear weapons at them in a first strike; because of this, and perhaps because of a sense of insecurity and paranoia with roots reaching back to the invasions of Russia by Napoleon and Hitler, they had aimed a huge arsenal of nuclear weapons at us.[40]

President Reagan, stirred up by the Committee on the Present Danger, dialed up the US nuclear rhetoric and posture, never believing the US would attack first. Nevertheless, Reagan's actions (SDI, nuclear deployments in NATO, war plans that included first-strike options, etc.) convinced Moscow that Washington was preparing for nuclear war. This misperception served to increase the risk of such a catastrophe. The Soviets were watching for signs of an attack, might have wrongly concluded one was imminent, and could have launched their own preemptive attack.

Russian leaders observed US nuclear activities that, to them, appeared to be preparations for a nuclear first strike. Yet President Reagan had no intention of launching one. Was Washington also misreading the tea leaves, seeing first-strike threats from Russia that did not, in fact, exist?

Yes.

## THE WALL COMES DOWN

It's one thing to debate Soviet nuclear intentions during the Cold War, when there was little information to go on. It was all too easy for politicians to hype the nuclear threat for their own purposes, and for defense hawks and conservative academics to interpret the inconclusive intelligence as they wished.

It is quite another thing, however, to continue to argue today, long after the Cold War has thawed, that Russia would consider a nuclear first strike now. One of the wonders of engagement with Russia is we no longer have to guess at such things; we can just ask them—or, better yet, read their archives.

As historian David Holloway wrote in 1994,

> It was not possible to analyze Soviet policy—as one would study American or British policy—in terms of the interplay of individuals, institutions and circumstances. Soviet nuclear weapons policy was often presented therefore as the product of the Soviet system, or of Marxist-Leninist ideology, or of an individual leader's policy goals. Only now, with the end of the Cold War and the collapse of the Soviet Union, is it becoming possible to write differently about Soviet nuclear weapons policy, to place it more securely in the context of Soviet history and the history of the Cold War.[41]

For example, as Holloway explains, Stalin had predicted another world war within twenty years after World War II, and that this massive conflict would see socialism defeat capitalism. But the advent of nuclear weapons was a thorn in the side of Soviet ideology. If nuclear war came, how would the Soviet Union survive and prevail? At first, the Soviet military regarded nuclear weapons as instruments of war and suggested that preemptive strikes and the Soviet Union's large size might give it an advantage over the West. But such thinking did not last long.

It was difficult for Soviet policy to make dramatic shifts until Stalin's death in 1953. But soon thereafter the term *peaceful coexistence* began to appear in Soviet statements. By asserting that capitalism and socialism could coexist for a long time, new Soviet leaders were rejecting Stalin's vision of inevitable war—and creating an alternative to nuclear holocaust.

Why this shift? Russia had developed a clear sense of how devastating nuclear war could be. The Soviets tested their first hydrogen bomb in 1953, and by then Russian scientists had concluded that the use of about a hundred large hydrogen bombs would "create on the whole globe conditions impossible for life." They found that "one cannot but acknowledge that over the human race there hangs the threat of an end to all life on earth."[42]

Nikita Khrushchev, who took over after Stalin, received a briefing on the dangers of atomic weapons in 1953. Many years later he said when he "learned all the facts about nuclear power I couldn't sleep for several days. Then I became convinced that we could never possibly use these weapons, and when I realized that I was able to sleep again. But all the same we must be prepared. Our understanding is not a sufficient answer to the arrogance of the imperialists."[43]

This construct explains a lot about US and Russian nuclear policy then and now: Do not use the bomb but, to deter the adversary, be prepared to use it at all times. As Khrushchev said in 1953 after Russia's first hydrogen bomb test, "Let these bombs lie, let them get on the nerves of those who would like to unleash war. Let them know that war cannot be unleashed, because if you start a war, you will get the proper response."[44]

Holloway writes, reflecting Eisenhower's view, that even by 1955 "a kind of existential deterrence had come into being" in which leaders on both sides "understood how terrible a nuclear war would be, and each side believed that the other understood this too. On this basis they shared the conviction that neither would start a nuclear war."[45]

This conviction did not stop presidents from developing weapons and policies to appear "tough" in the eyes of Russian leaders or American voters, or to seek a sense of assured "victory" if war happened. It did not stop the arms race, the pursuit of military advantage, or the use of nuclear threats. Nevertheless, the common United States–Soviet understanding that nuclear war was to be avoided at all costs had become a basic premise in the Cold War rivalry.[46]

## "A DIALOGUE OF THE EYES"

We would have to wait thirty long years for the United States and the Soviet Union to formally agree that neither would start a nuclear war. President Reagan and Premier Gorbachev met for the first time in Geneva on November 19, 1985. Together, the leaders controlled about 60,000 nuclear weapons at the time, and the arms race was in full swing.

The two leaders faced off, lacking a common language but managing to exchange information, in a way. As Gorbachev recalled, "We extended a hand to each other, and started talking. He speaks English, I speak Russian, he understands nothing, I understand nothing. But it seems there is a kind of dialogue being connected, a dialogue of the eyes."

At the summit, the two agreed that "nuclear war cannot be won and must never be fought."

"Can you imagine what that meant?" Gorbachev told veteran journalist David Hoffman. "It meant that everything we had been doing was an error. Both of us knew better than anyone else the kinds of weapons that we had. And those were really piles, mountains of nuclear weapons. A war could not start because of a political decision, but just because of some technical failure."[47]

Once both nations agreed they would not attack the other, the logic of Cold War nuclear policy began to crumble. If there is no significant risk of a disarming first strike, then there is no need to launch nuclear weapons first, preemptively or quickly; no need for presidential sole authority other than for retaliation; no need for weapons on high alert; no need to launch weapons on warning of attack; and, indeed, no need for ground-based ballistic missiles at all.

After the summit, Secretary of State George Shultz began to press this point. "The Soviets," he told Reagan, "contrary to the Defense Department and the CIA line, are not an omnipotent, omnipresent power gaining ground and threatening to wipe us out."[48] Secretary Shultz wanted Reagan to consider trading limits on SDI for cuts in ballistic missiles. But Defense Secretary Caspar Weinberger opposed any limits on SDI. Then, in June 1986, Weinberger made a radical proposal: the United States and the Soviet Union should eliminate all ballistic missiles. "Everyone was astonished," Shultz recalled.

But such hopes were dashed at the Reykjavik Summit in October 1986. The talks came tantalizingly close to eliminating all nuclear weapons but were blocked by Reagan's misplaced belief in a national missile defense system. Gorbachev wanted missile defense research to be limited to laboratory experiments, and Reagan refused.

Even so, Gorbachev and Reagan closed a dangerous chapter in the Cold War. The two leaders made it clear to all that neither nation had any intention of attacking the other with nuclear weapons. In fact, they had been open to eliminating all of them.

This diplomatic momentum was strong enough to achieve the Intermediate-Range Nuclear Forces (INF) Treaty in 1987, which eliminated an entire class of mid-range nuclear and conventional missiles from Europe, the first agreement to actually reduce nuclear weapons. This is the same agreement from which President Trump withdrew in August 2019.

Then in 1991, President George H. W. Bush and Gorbachev signed the START Treaty, which reduced strategic offensive nuclear forces by about 30 percent. This agreement set a common limit on both sides' strategic (long-range) nuclear forces and represented a shift away from either side seeking a strategic advantage.

The further we get from the Cold War, the more implausible an unprovoked Russian nuclear attack becomes. President Obama, according to his senior nuclear policy advisor Jon Wolfsthal, did not think that the United States or Russia would ever initiate a nuclear attack. "Any of the exercises we did, any of the discussions we had, any of the intel that we went through, none of them ever turned on a Monday-everything's-fine-Tuesday-Russia-wakes-up-and-decides-let's-go and vice versa," said Wolfsthal in an interview.[49] He said that "the president and vice president both clearly viewed the overkill, the oversystemization, the time pressures, and the process to be anachronistic and detached from real-world realities."

Asked if he thought Russia was planning a first strike, Wolfsthal said, "Absolutely not. My belief is the Russians don't even plan or exercise for a bolt out of the blue, that it never occurs to them because even if they were to be successful it would still destroy everything that they value in terms of kleptocracy, oligarchy, control over their country, the global economy—all that goes away. They know it as well as we do."

Even so, the core nuclear policies from the Cold War have been left unchanged: first use, sole authority, launch on warning, ballistic missiles on high alert. Everything has changed, and yet nothing has. The Cold War and the Soviet Union are long gone, but the US president can still unilaterally launch a nuclear war in about the time it takes to order a pizza.

Why? Even though most officials in the nuclear establishment do not believe Russia would launch a first strike, it still has the capability to do so. "I do not think that the Russians intend to launch a no-notice, massive nuclear strike on the United States," then STRATCOM Commander Kehler said in 2019. "But they have the capability to do it. And as long as they do, my view is we have got to be able to respond to that kind of an attack quickly, if that's the decision that we need to make."[50]

US nuclear plans are still based on Russian *capability*, which is based on US capability, even if leaders in Moscow and Washington have no intention of using that capability. This "conservative" approach fails to recognize that by maintaining our unneeded ability for quick nuclear response we are increasing the risk of starting a nuclear war by *accident*. These planners think they are playing it safe, but, in reality, they are perpetuating tremendous risks for all of us, as we will see in the next chapter.

# CHAPTER 3

# BLUNDERING INTO NUCLEAR WAR

*We live in a time where a fateful error or miscalculation, rather than
an intentional act, is the most likely catalyst to nuclear catastrophe.*
—JOAN ROHLFING, PRESIDENT, NUCLEAR THREAT INITIATIVE[1]

January 13, 2018, started out like any other Saturday on Hawaii—with
a beautiful sunrise. But at 8:07 a.m. the day took a dramatic turn.
The Hawaii Emergency Management Agency broadcast an official message to more than one million people: "Ballistic missile threat inbound to
Hawaii. Seek immediate shelter. This is not a drill." This was just days after
Kim Jong-un said his nuclear launch button was "always on my table" and

President Trump tweeted that his button was "much bigger & more powerful," so a no-warning nuclear attack from North Korea seemed plausible.

Hawaiians panicked and ran through the streets in terror and confusion. Parents opened manhole covers and pushed their crying children down into the sewers for protection. People with relatives in different locations struggled to decide who to go to first. In stores, customers lay down on the floor beside strangers. Terrified drivers sped at 90 miles an hour, racing to find shelter or loved ones.

How long until the missiles arrive? Where will they land? Are they carrying nuclear warheads? No one knew. No one was prepared for this.

Desperate Hawaiians waited thirty-eight long minutes for a nuclear attack that never came. It was a false alarm. A Hawaii Emergency Management Agency employee, who "believed that the missile threat was real," pushed the wrong button, and the agency "didn't have reasonable safeguards in place to prevent human error from resulting in the transmission of a false alert," according to an investigation by the Federal Communications Commission.[2]

"None of us can find a satisfying way to explain to our children why we still have nuclear weapons, much less allay their fears," wrote Cynthia Lazaroff, who was on the island of Kauai at the time.[3]

"This threat of nuclear war, nuclear attack, is not a game. This is real. And this is what the people of Hawaii just went through," said Representative Tulsi Gabbard (D-HI).[4]

---

The Hawaii false alarm was a wake-up call. Machines can malfunction, and people, being human, will err. Had the Hawaii alert been part of the national warning system, the president would have been alerted and this could have led to a nuclear response.

Hawaii "drove home the fact that a false alarm could happen in the central nuclear system," said Bruce Blair, a former Air Force missile launch officer. "It just reminds everyone that these systems are susceptible to human and technical error. We have false alarms, and we have more of them now than we ever have had, even during the Cold War."[5]

Despite the fact that neither the United States nor Russia would intentionally start a nuclear war, the launch policies on both sides support the option to strike first, and thus both nations are prepared to respond quickly to a surprise attack. But preparing for a quick response creates more problems than it solves. The real danger today is nuclear war starting from mistakes and false alarms as happened in Hawaii, *not* an intentional Russian attack against the United States.

Like a runner at the starting line waiting for the shot to be fired, we are bracing ourselves for a surprise attack that is extremely remote. Shadowboxing with a danger that is not really there, we tragically make ourselves more vulnerable to the real threat—that we might blunder into a nuclear catastrophe. Preparing for a disarming first strike leads us to take on dangerous policies: giving the president sole authority to launch, preserving the option to launch first, keeping the weapons on high alert, and preparing to launch on warning of attack. These options inextricably make a nuclear blunder more likely.

A nuclear war by accident or mistake would be just as deadly as one by intent. The size and lethality of US and Russian forces ensures that a nuclear exchange—regardless of *why* it started—could result in the end of our civilization.

We are not simply raising a *theoretical* possibility. We came, in fact, very close to blundering into nuclear catastrophe several times during the Cold War. Gorbachev kept a sculpture of a goose in his Moscow office to remind him that US radars had once mistaken a flock of geese for a Soviet bomber attack.[6]

Of all the ways we brace for a surprise attack, giving the president sole authority to launch is perhaps the most dangerous. As in the scenario envisioned in the preface, sole authority combined with first use and alert forces means that a president could act independently and swiftly. Launch on warning creates time pressures under which no president could make a fully informed choice. By eliminating sole authority for first use, we would be acknowledging that we can take our time to make a retaliatory launch decision. With no significant risk of a surprise attack, we don't need forces on alert that could be launched on warning. And sharing nuclear decision authority with a democratically elected group, such as Congress, would slow the process down and ensure that the first use of nuclear weapons would be vigorously debated. Sole authority is the place to start.

There are three main ways that presidential sole authority increases the risk of nuclear conflict. First, the president could be working with incomplete information that, if acted on, would lead to catastrophe. The pressure to make a launch decision in about ten minutes or less is incompatible with the need to verify the information that is driving the decision to launch. "But everything would happen so fast that I wondered how much planning or reason could be applied in such a crisis," President Reagan said. "The Russians sometimes kept submarines off our East Coast with nuclear missiles that could turn the White House into a pile of radioactive rubble within six to eight minutes. *Six minutes* to decide how to respond to a blip on a radar scope and decide whether to release Armageddon! How can anyone apply reason at a time like that?"[7]

Second, the president could be emotionally unstable or under the influence of drugs or alcohol and could impulsively choose to initiate nuclear war at any time. Third, the president could launch nuclear weapons based on information generated by a technical miscalculation—a false alarm, for example—or a cyberattack.

In any of these cases, even if the president's advisors recognize the problem, they have no authority to override the president's decision. It is really true that the president has the "nuclear button," and no one is authorized to stop her or him from using it. This is an unacceptable risk that can be eliminated only by putting constraints on the president's authority to start a nuclear war.

"It is correct to say that no well-intentioned, coolly rational political or military leader is likely to initiate the use of nuclear weapons," wrote McNamara. "But political and military leaders, in moments of severe crisis, are likely to be neither well informed nor coolly rational."[8]

## BAD INFORMATION

Presidential mistakes regarding the bomb go back to its earliest days. In 1945, as President Truman was debating how to use the first atomic weapons, he apparently did not realize that Hiroshima was a city and not just a military site. Truman wrote in his journal on July 25, just two weeks before bombing Japan, that the bomb's first target was a "purely military one" and that "soldiers and sailors," and "not women and children," would be the victims. This is clearly wrong; while Hiroshima contained an army headquarters, it was also a city populated mainly by civilians (not soldiers), who made up the vast majority of the victims. Truman's reference to Hiroshima as "purely military" stands as a shocking misunderstanding that may have led to tens of thousands of unintended civilian deaths. Truman's realization of this mistake may have contributed to his decision to halt the bombings after Nagasaki, claiming that he didn't like the idea of killing "all those kids."[9]

The 1962 Cuban Missile Crisis began with the discovery that the Soviet Union had shipped nuclear-capable ballistic missiles onto the island. Even though, at the time, the United States had a significantly larger nuclear arsenal than Russia, the Kennedy administration did not believe this numerical superiority translated into an ability to launch a successful first strike against Russia. This force discrepancy may, however, have led the Soviets to worry that the United States might *think* it had a significant strategic advantage. This fear may have led Soviet leader Khrushchev to place weapons in Cuba that could reach the United States.

The Kennedy administration did not see the new missiles in Cuba as changing the military balance—both sides were already deterred from attacking each other with nuclear weapons. But the missiles did present a political challenge that had to be addressed—and reversed.

The majority of Kennedy's senior advisors advocated for a conventional air attack to destroy the missiles in Cuba. Such an attack would likely require a follow-up ground attack, which in turn had the danger of escalating into a nuclear conflict. Indeed, we now know, which we did not know then, that the Soviets had *tactical* (short-range) nuclear weapons in Cuba that would have been used to repel a ground attack. Defense Secretary Robert McNamara recommended a sea blockade to prevent Soviet supplies from reaching Cuba until the missiles were removed. Fortunately, the president chose option two.

During the Cuban Missile Crisis, I (Bill) was managing an electronic defense laboratory in California but also served as a pro bono CIA consultant. When a U-2 flight revealed the Soviet missile deployment in Cuba, I was called by the deputy CIA director, who asked me to come to Washington immediately to consult with him on a matter of great national importance. When I arrived in his office the next morning, he showed me the U-2 pictures and I knew immediately we were facing a grave security crisis that could well end up in a catastrophic nuclear war; I offered to do anything I could to help.

He told me that he wanted me to work with a small team that would analyze the incoming intelligence data and prepare a report each morning for the president telling him the status of the Soviet nuclear deployment. President Kennedy had decided against a military attack in favor of diplomacy but was prepared to abandon diplomacy just before the missiles became operational. The big question was: How much time did he have for diplomacy? And he wanted that answer updated every morning. The job of our team was to give him as many days as possible for his diplomacy. In the meantime, as we could see each day, the Soviet missile crews were working desperately to bring their missiles into operational status. The situation was grave, and I believed that there was a real risk of all-out United States–USSR nuclear war. Indeed, every day that I went into the analysis center I believed would be my last day on earth.

We were spared that catastrophe by the determination of the two leaders to avoid it, and by their creative diplomacy. But all of the evidence argues that in spite of their determination and their creativity, they almost failed. As Khrushchev put it in a message to Kennedy during the crisis: "If war should break out, it would not be in our power to stop it—war ends when it has rolled through cities and villages, everywhere sowing death and destruction."[10]

President Kennedy later said that he thought the Cuban Missile Crisis had a one-in-three chance of ending in a nuclear war. Those are very high odds for what would effectively be the end of civilization. But, in fact, the odds of a catastrophe were even higher. When Kennedy made that estimate he did not know that Soviet forces in Cuba already had completed the operational readiness of about a hundred tactical nuclear weapons and had the authority to use them in self-defense. The US Joint Chiefs of Staff had recommended a conventional military attack on Cuba based on their belief that the Soviets did *not yet* have operational nuclear weapons. If Kennedy had accepted that recommendation, US troops would have been decimated on the beachhead with tactical nuclear weapons and a general nuclear war would surely have followed. The world avoided a nuclear holocaust over Cuba as much by *good luck* as by good management.

"The fog of fear, confusion and misinformation that enveloped the principals caught up in the Cuban Missile Crisis could have at any moment led to nuclear annihilation," wrote former STRATCOM commander Gen. George Lee Butler. "The chilling fact is that American decision makers did not know then, and not for many years thereafter, that even as they contemplated an invasion some one hundred Soviet tactical nuclear warheads were already in place on the island. No further indictment is required to put elegant theories of nuclear deterrence in perpetual question."[11]

When Robert McNamara learned about this in 1992, thirty years later, he said: "We don't need to speculate what would have happened. It would have been an *absolute disaster* for the world . . . No one should believe that a US force could have been attacked by tactical nuclear warheads without responding with nuclear warheads. And where would it have ended? In utter disaster."[12]

The point of this horrifying story is that even presidents, who have the benefit of their various and expansive intelligence services, often do not have complete information. And what they don't know might be crucial. In this case, Kennedy was not yet considering the use of nuclear weapons. But he was making serious decisions about using military force, and he did not know the facts on the ground. In a similarly tense situation where the use of atomic weapons was under consideration, delaying that decision and consulting with a wider circle of people would be critical. Sole authority in such a situation would be dangerous.

Kennedy was misinformed about nuclear weapons in Cuba, but he also could not have known how Khrushchev was thinking about his options. We now know that Khrushchev was in fact seeking a way out, even while his military advisors were pushing him to not back down:

> When I asked the military advisors if they could reassure me that hold-ing fast would not result in the death of five hundred million human beings, they looked at me as though I was out of my mind, or what was

worse, a traitor. The biggest tragedy, as they saw it, was not that our country might be devastated and everything lost, but that the Chinese or the Albanians might accuse us of appeasement or weakness.

So I said to myself, "To hell with these maniacs. If I can get the United States to assure me that it will not attempt to overthrow the Cuban government, I will remove the missiles." That is what happened, and now I am reviled by the Chinese and the Albanians . . .

They say I was afraid to stand up to a paper tiger. It is all such nonsense. What good would it have done me in the last hour of my life to know that though our great nation and the United States were in complete ruins, the national honor of the Soviet Union was intact?[13]

Happily, we do not face a situation like the Cuban Missile Crisis today, but it certainly is possible to imagine a crisis on the Korean Peninsula where the United States decides to use conventional military power against North Korea, based on the calculation that this would not escalate to a nuclear war. But North Korea built its nuclear arsenal to keep its regime in power. If its leaders saw a conventional American attack as threatening the survivability of the regime, they might well use nuclear weapons in a desperate attempt to maintain their power, and if they did, that political miscalculation could lead to millions of casualties on all sides. And if China or Russia were to intervene, it could lead to a nuclear holocaust.

Another danger of sole authority is that presidents, particularly since the end of the Cold War, do not spend enough time learning about nuclear weapons and what their options would be. Thus, if a crisis were to occur, the president may be acting on misinformation and would be unprepared to handle the huge responsibility of sole authority. Adm. James Winnefeld, a former commander of North American Aerospace Defense Command (NORAD) and vice chairman of the Joint Chiefs of Staff, wrote that "any president will likely not possess a deep understanding of the system that will underpin the lowest probability but highest consequence decision anyone on the planet will ever make."[14]

Even if presidents are wise enough to turn to their staff for advice, they may become overly reliant on the person who would brief them. What if the

STRATCOM commander, the most likely briefer, is also badly informed or having a bad day?

Jon Wolfsthal served as President Obama's special assistant for national security affairs and senior director at the National Security Council for arms control and nonproliferation. He felt confident that Vice President Joe Biden and STRATCOM commander Gen. Jim Cartwright had a deep understanding on nuclear issues to assist Obama. In an interview for this book, Wolfsthal said, "They worked so closely with the president on so many other issues that there was a level of trust and understanding that they could say, 'Mr. President, I just want to tell you before you make this decision that these are your other options and you might consider them if you have different goals here.'" But President Trump brings the dangers of sole authority into focus because "we've assumed that we're going to have a thoughtful, caring, dedicated, smart actor in that position who has a relationship of trust with his subordinates, and that's not what we have with Trump," Wolfsthal said.

Wolfsthal said that "the men and women of STRATCOM are dedicated, thoughtful, intelligent, and very aware of their awesome responsibility." But he added that the manner in which the commander of STRATCOM briefs the president on her or his options matters a lot. "Some are really good at it, and some are not. I've seen mistakes made in briefings, even by top-level officials, that shouldn't happen. The system is far from perfect, just like people are far from perfect."

Wolfsthal continued, "People in the United States, and I'm sure this is true in Russia, believe that the risks associated and inherent with having nuclear weapons and threatening their use are totally manageable. That we're totally up to the challenge because we take it seriously and we will be able to effectively manage it, despite the fact that every couple of years we see instances where that is patently not true, and where it's not true in any other aspect of our lives—doctors make mistakes and people die; good drivers drive off the road because they're texting and kill somebody. It happens in every walk of life. Why wouldn't it happen in nuclear? And, of course, it does."

"To me, the big, missing component of sole authority," Wolfsthal said, "is the human predilection to mistakes and how that gets accepted and baked in and dismissed when it comes to all of these questions: nuclear command

and control, first use, action in a crisis, communicating things down the line—there's always a risk factor there and it scares me."[15]

## UNSTABLE PRESIDENTS

President Trump is not the first American leader who raises concerns about sole authority. As mentioned in chapter 1, President Kennedy's medical records reveal that he suffered from more ailments, was in far greater pain, and was taking many more medications than the public knew at the time.[16] President Reagan was formally diagnosed with Alzheimer's in 1994, after he had left the White House, but he displayed some early symptoms while he was still in office.[17] Both Kennedy and Reagan's mental states while they were in office could have been affected by their ailments and the medications they took.

Besides Trump, no president of the atomic age has caused more concern than Richard Nixon and his tendency to drink alcohol to excess. In 1969, a US spy plane was downed by North Korea over the Sea of Japan, killing thirty-one Americans. "We were being tested," Nixon wrote in his memoirs, "and therefore force must be met with force."[18]

George Carver, the CIA's top Vietnam specialist at the time, recalls that "Nixon became incensed and ordered a tactical nuclear strike . . . The Joint Chiefs were alerted and asked to recommend targets, but [national security advisor Henry] Kissinger got on the phone to them. They agreed not to do anything until Nixon sobered up in the morning."[19]

Then, on October 20, 1973, Nixon was reportedly drunk and unable to respond when the Arab–Israeli conflict reached the brink of nuclear war. Syria and Egypt, in an effort to regain territories lost in the Six-Day War, launched a joint surprise attack, and Israel and the United States were caught off-guard. The Soviets became involved, and it wasn't long before they threatened military action.[20]

It was later revealed that US intelligence believed that Soviet ships were carrying nuclear arms through the strait that connects the Black Sea to the Mediterranean, on their way to Egypt. This assessment is now widely disputed, but at the time Washington responded by placing its nuclear forces on a global alert for only the fourth time in history. [21]

In the midst of this crisis Kissinger (then secretary of state) reported on a call he had received from his aide, Brent Scowcroft:

> SCOWCROFT: The switchboard just got a call from 10 Downing Street to inquire whether the president would be available for a call within thirty minutes from the prime minister. The subject would be the Middle East.
>
> KISSINGER: Can we tell them no? When I talked to the president he was loaded.

Kissinger, Defense Secretary James Schlesinger, Joint Chiefs Chairman Adm. Thomas Moorer, White House Chief of Staff Alexander Haig, and CIA Director Bill Colby took action to warn the Soviets to back off. Nixon, as far as we know, never took part in the negotiations.

Larry Eagleburger, who was a National Security Council staffer under Kissinger during the Nixon administration and later became secretary of state under President George H. W. Bush, noted, "One of the things that I recall now with a great deal more equanimity than I did at the time is what was never really understood: the degree to which the Watergate crisis, particularly in its final months, meant that if we had been put to the test somewhere in the foreign policy arena, we would not have been able to respond. We were a ship dead in the water."[22]

Defense Secretary James Schlesinger was so concerned about Nixon that he reportedly told military commanders that if the president ordered a nuclear launch, they should check with him or Secretary of State Kissinger first. Senator Alan Cranston phoned Schlesinger, warning about "the need for keeping a berserk president from plunging us into a holocaust." I (Bill), in a private discussion, once asked Schlesinger about this report. He did not confirm it. But he had an excellent opportunity to deny it and did not.

President Trump does not, as far as we know, drink, take pain medication, or suffer from dementia. But his lack of attention to detail, uneven temperament, erratic behavior, and domestic political troubles have led us to be concerned about his judgment and fitness to control the nuclear arsenal. But to be precise, our concern over sole use and first use is not just about President Trump. No president should have to make that awesome decision quickly, without deliberation and consultation; no president should have the sole authority to launch nuclear weapons first; no single person should have the power to end the world. "It makes no sense to me that we would vest such a tremendous and grave responsibility in a single human being," said Joan Rohlfing, president of the Nuclear Threat Initiative.[23]

Finally, even a well-informed, rational president may lose control of the football to someone who is not. In August 1991, Gorbachev was taking a working vacation in the Crimea, reviewing a draft treaty to decentralize the Soviet Union by giving the republics greater independence, such as control of their own resources. The *chemodanchik* was with him, as always. The Soviet command system required three leaders—the president, the defense minister, and the chief of the General Staff—to give permission to launch, and all three had their own footballs. Once authorized, the launch order would be disseminated by the General Staff to the commanders of the armed forces—air, rocket forces, and navy. Gorbachev "regarded the whole process with disdain" and "abhorred the thought of nuclear war," according to Hoffman.[24]

At 4:32 p.m. on Sunday, August 18, all communication links at Gorbachev's compound were dead. Soviet nuclear forces were disconnected from their civilian commander. A high-level delegation arrived unexpectedly from Moscow demanding Gorbachev turn over power to Vice President Gennady Yanayev. Gorbachev refused. The delegation left, but for the next three days Gorbachev and his family were prisoners in their own vacation compound. Back in Moscow, the coup plotters announced that Gorbachev was ill and had been replaced by Yanayev. Boris Yeltsin, president of the Russian republic, the largest of the newly independent Soviet states, was calling on Russians to resist the coup.

On Monday morning, back in Moscow, the General Staff ordered senior officers to get Gorbachev's football back from him, which they did. It was flown back to Moscow on Gorbachev's plane.

The coup collapsed on Wednesday, August 21. Gorbachev had lost control of his nuclear football, but at least one of the three military officials required to launch an attack, Air Force Gen. Yevgeny Shaposhnikov, had opposed the coup, and it is possible that the heads of the rocket forces and the naval forces did as well. Gorbachev got his football back, but the fact that he lost it at all is a chilling reminder that even well-intentioned plans can go wrong.

## FALSE ALARMS

Turning back to the United States, we have more to fear from a technical miscalculation than political uprising. In addition to concerns about bad information and bad judgment, the president could launch nuclear weapons because of an error in our systems—by responding to a false alarm, such as happened in Hawaii. The United States had at least three such false alarms during the Cold War, and the Soviet Union had two that we know about. And there is no reason to think that they could not happen again.

To understand how false alarms can happen, it is useful to take a quick tour of the US system that provides warning of possible missile attacks. The US Missile Attack Warning System has three major segments: sensors (satellites and radar) to detect missile launches, computer centers to analyze and distribute sensor data, and command posts to assess information and take actions. The major command posts include the North American Aerospace Defense Command (NORAD) at Colorado Springs, Colorado; Strategic Command (STRATCOM) at Offutt Air Force Base, Nebraska; the National Military Command Center (NMCC) at the Pentagon; and the Alternate National Military Command Center (ANMCC) at Fort Ritchie, Maryland.

If Russia were to launch a nuclear attack, here is what the US warning system would see: First, infrared satellites would detect the heat from burning missile motors as they lift nuclear warheads into space. Next, ground-based radar would track the missiles in flight. Some radar (called Ballistic Missile Early Warning System, BMEWS) would detect Soviet ICBMs, while others (called Pave Paws) would track submarine-launched missiles.

The system generates many false positives. Before ICBMs, radar had been fooled by flocks of high-flying geese into warning that Soviet bombers were attacking the United States over the North Pole.[25] A bomber warning allows hours to determine whether the threat is real and to get planes in the air, and those planes can later be recalled. But if the alert involves ballistic missiles, a decision to launch ICBMs must be made in under ten minutes, and they cannot be recalled.

Why can't nuclear-armed ballistic missiles be recalled or aborted? It is a reasonable question. Test missiles, without warheads, are armed with self-destruct mechanisms and can be blown up if the missile malfunctions or strays off course. But, in the case of real launches with armed missiles, the Air Force is worried that any kind of radio-controlled self-destruct or "kill switch" could be hacked by an adversary, who would then be able to disarm the entire missile force.

As retold by Eric Schlosser in his book *Command and Control* and by Dan Ellsberg in *The Doomsday Machine*, the first BMEWS radar at Thule Air Base, Greenland, came on line in 1960 and had problems right away. Industry officials were being given a tour of NORAD when the computers warned that the United States was under attack. The executives were escorted out and left to believe that nuclear war was about to begin.

Although there was a sense of panic at NORAD, it was probably reassuring that, as it happened, Soviet leader Khrushchev was at the United Nations in New York that week. The false alarm had been caused by radar signals bouncing off the moon as it rose over Norway.[26]

From January 1979 through June 1980, there were 3,703 "routine missile display conferences" to review ambiguous missile launches. If NORAD's watch officers identify a potential threat, they convene a "threat assessment conference" with more senior officials, such as the chairman of the Joint Chiefs of Staff. Four such conferences were held during this time. If the threat persists, NORAD calls a "missile attack conference," which includes the president. No such conferences were called then, but we came close.[27]

The first threat assessment conference was easily resolved; an unarmed rocket was misidentified as a missile. But the second, on November 9, 1979, was a doozy. The US early warning system was flashing red. The computers at NORAD were saying that the United States was under nuclear attack. Soviet missiles had been launched from submarines off the West Coast. Then more missiles appeared on the computer screens, now launched from inside Russia. It appeared to be a major surprise attack.

But global tensions were not particularly high, and a real attack seemed unlikely. Watch officers at NORAD, buried deep underground the Cheyenne Mountain complex in Colorado, checked in with the radar, satellites, and other sensors that were supposedly reporting the attack. Nothing there. The computers were seeing an attack, but the sensors were not.

Following procedure, bomber crews at Strategic Air Command (SAC, now STRATCOM) bases were ordered to their planes, missile crews were put on alert, and fighter planes were scrambled (sent aloft as quickly as possible).

But the attack never arrived. It was a false alarm. In this case, the watch officers may have been more open to interpreting the attack as false, considering that this would have been a suicidal act on the part of Russian leaders and that we were not in the midst of an international crisis. But what if this event had happened during the Cuban Missile Crisis, or a Mideast war? Had the watch officers come to a different conclusion, the alert could have gone all the way to the president, waking him, and giving him perhaps ten minutes to make a fate-of-the-world decision with little context or background to inform that choice.

What really happened? It was first reported that a technician had mistakenly put military exercise tapes into the Pentagon's computer system, which sent realistic details of a war game to SAC and the Pentagon. That narrative is still the conventional wisdom today.

But, in fact, this was no simple operator error, and the truth is more concerning. It turned out that software simulating a Soviet missile attack in Pentagon computers was *inexplicably* transferred into the regular warning display at NORAD's headquarters. According to General Accounting Office testimony, the NORAD computer, "failing to recognize test data being used for software development, generated inappropriate warning of a massive Soviet missile attack." Indeed, NORAD later acknowledged that the "precise mode of failure . . . could not be replicated," and NORAD could not determine whether the cause was "human error, computer error, or a combination of both."[28] The United States had likely come within minutes of launching a massive nuclear attack by mistake, and the cause was never fully understood.

The Soviets took note of the event and expressed their justified concerns. Premier Brezhnev sent a message to President Carter stating that the situation was "fraught with a tremendous danger" and that "I think you will agree that there should be no errors in such matters."[29]

## THE 46-CENT CHIP

The most realistic false alarm took place on June 3, 1980. In the middle of the night, about 2:30 a.m. Eastern Time, National Security Advisor Zbigniew Brzezinski received a phone call from William Odom, a military assistant. Odom informed Brzezinski that Soviet submarines had launched 220 missiles against the United States. The Soviet Union had recently invaded Afghanistan, making a real attack more plausible. Brzezinski believed that he would have three minutes to confirm the attack, and then President Carter would have four minutes to make a decision. That was not much time to make a truly existential choice.

Brzezinski told Odom he would wait for a call to confirm the Soviet launch and the intended targets before calling the president. Brzezinski was convinced the United States should launch a retaliatory strike and told Odom to confirm that Strategic Air Command was launching its nuclear-armed bombers. When Odom called back, things had gotten worse. It was now reported that the Soviets had launched 2,200 missiles, an all-out attack. The United States would be utterly destroyed, no matter what Brzezinski did. He decided not to awaken his wife, thinking that everyone would be dead in half an hour. Just before Brzezinski intended to call the president, Odom called a third time to say that other warning systems were not reporting an attack. It was a false alarm.[30]

I (Bill), as undersecretary of defense at the time, had a personal role in this drama. After the alarm was found to be false, the NORAD watch officer called me. The general got right to the point: he told me that his warning computer was showing hundreds of ICBMs missiles in flight from the Soviet Union to targets in the United States. For one heart-stopping moment I thought my worst nuclear nightmare had come true. But the general quickly explained that he had already concluded this was a false alarm—he was calling to see if I could help him determine what had gone wrong with his computer!

Then, just a few days later on June 6, NORAD computers warned the Pentagon of a third attack. Once again, sirens wailed and bomber crews were sent running to their planes, but it was again found to be a false alarm. This time technicians found a defective computer chip in a communications device. NORAD sent test messages to ensure that it could talk to computers at the Pentagon and elsewhere. The test message warned of a missile attack and indicated the number of missiles in flight. In a test, the number should be zero. But the bad computer chip put a number 2 in that space, sending the erroneous message that 220 or 2,200 missiles had been launched. The bad chip was replaced, at a cost of 46 cents.[31] Future test messages did not include the number of missiles.

In a July 17, 1980, memo to President Carter, Defense Secretary Harold Brown concluded that "I believe we must be prepared for the possibility that another, unrelated malfunction may someday generate another false alert."[32]

An October 1980 report by the Senate Armed Services Committee found that "there is no guarantee that false alerts will not happen in the future. They will occur and we must rely on the collective judgment of the people manning the system to recognize and deal correctly with false alarms."[33]

A 1981 General Accounting Office report on the false alarm incidents found that NORAD's computer problems "not only attracted adverse publicity but nearly caused an international crisis." The report also stated that "the erroneous indications of launches could occur again because no system can be built to cover all exceptions." As Congressman Jack Brooks, chairman of the House Committee of Government Relations, put it at the time, "These events are most alarming."[34]

Earlier in this chapter, I (Bill) told about my personal experience with a false alarm. That experience made an indelible impression on me. Since then, I have never regarded false alarms as an academic problem. They have happened in the United States and Russia; they can happen again. It is the nature of life: humans do err; machines do malfunction. Because of the safeguards we have built into the system, I believe that there is a low probability that a false alarm would lead to an accidental war. But a low probability is not zero.

An accidental nuclear war is unlikely, yet its consequences would threaten the very existence of our civilization, so we should not accept even low probabilities of a false alarm if we do not have to. And we do not. We can modify our nuclear policies and our nuclear forces to eliminate the risk of a false alarm that could lead to catastrophic results (see chapter 10).

## SOVIET FALSE ALARMS

On September 26, 1983, a Soviet early-warning satellite indicated five US nuclear missile launches. Tensions were already high because the Soviet Union had mistakenly downed a South Korean passenger plane just weeks before. Lieutenant Colonel Stanislav Petrov, the watch officer on duty, had just minutes to decide what to do.

Operating mostly by instinct, Petrov decided this was a false alarm, reasoning that "when people start a war, they don't start it with only five missiles."[35] He was right.

Eventually it was determined that the satellites were fooled by sunlight reflecting off the tops of clouds. Soviet satellites were designed to minimize the chances of false alerts, but that night, soon after the equinox, satellites, sun, and US missile fields aligned in a way that no one had expected.

Petrov's quick thinking earned him the nickname "the man who saved the world."[36] If the satellite data had indicated a much larger attack, or if a different officer had been on duty, this false alarm could have turned into a catastrophe.

On January 25, 1995, a Russian early warning radar detected a missile launch off the coast of Norway. The missile's flight path appeared similar to that of a US submarine-launched ballistic missile, leading Russian officers to believe that the missile might detonate a nuclear warhead high in the atmosphere, blinding Russian radars before a larger attack. Russian nuclear forces went on full alert.[37]

But Russian early warning satellites did not detect a larger US attack and, fortunately, watch officers decided this was a false alarm. The detected missile was actually a US–Norwegian scientific rocket (the Black Brandt XII) studying the aurora borealis, or "northern lights." Norway had responsibly notified Moscow in advance of the launch, but the information didn't get into the right hands. A harmless science experiment had gone dangerously awry.[38]

President Boris Yeltsin put on a show for the media by activating his "football," the device used to authorize nuclear launches, even though he knew the alarm was false.[39]

## MORE ALARMS

During the Cold War, false alarms, fortunately, never made it as far as the US president. But under presidents Bush and Obama, the United States has, on multiple occasions, detected ambiguous ballistic missile threats that have led to the notification of the president and an exchange between the head of STRATCOM in Omaha and the president. According to Blair, the STRAT-COM commander is the only briefer in the conference, and on some of these occasions some of the president's key advisors did not get on the conference line in a timely way. "So, all I know is that during the GW Bush and the Obama years, there were events that involved the notification of the president in real time. And that never happened during the Cold War," said Blair.[40]

Trump may now have the same experience.

It is simply inexcusable, knowing that the system can fail, has failed, and indeed could fail again, that the United States still maintains weapons on high alert with the option of launch on warning. This is the best evidence we know of to prove the point that the United States has not bothered to learn the most important lessons of the Cold War. And by failing to learn these lessons, we may be doomed to repeat our most dangerous mistakes.

"Having successfully proposed to President Bush in 1991 to reduce bomber launch readiness from several minutes to several days," wrote Gen. Butler, "I am appalled that eight years later land- and sea-based missiles remain in what amounts to immediate launch postures. The risk of accidental or erroneous launch would evaporate in an operational environment where warheads and missiles are de-mated and preferably widely separated by location."[41]

The overriding danger of blundering into nuclear war is magnified by the US policy that gives the president clear and sole authority to launch a nuclear attack. The greatest danger of a nuclear exchange during the Cold War came *not* from a deliberately planned attack, but through bad information, unstable leadership, or false alarms. We have been focused on the wrong threat. And as we will see in the next chapter, concerns about cyberattacks make this problem even worse.

# CHAPTER 4

# HACKING
# THE BOMB

*If I were now in charge of the nation's security, I would try to make
sure we had played out every potential bad thing that could happen
with the conjunction of the nuclear situation as it now is, and the
possibility of effective cyberattacks to cause people to make mistakes.
Somebody could be intentionally trying to get us to make mistakes.*
—PRESIDENT BILL CLINTON[1]

n 2009, the United States and Iran were going head-to-head in a high-stakes
battle over Tehran's program to enrich uranium, a key bomb fuel. Iran's
uranium enrichment program was on track to install nine thousand centrifuges at its Natanz plant, about 150 miles south of Tehran. By the summer, Iran had produced enough low-enriched uranium, if enriched further,
to achieve an initial nuclear-weapons capability. If Iran continued production

at this rate, it would have had enough uranium for two nuclear weapons in a year's time.[2]

The United States and Israel had been seeking a way to stop Iran's program without launching an overt military strike, such as Israel's attacks against nuclear facilities in Iraq in 1981 and Syria in 2007. In 2009, Israel reportedly approached the George W. Bush administration for bunker-busting bombs and other equipment it would need for an air attack, telling the White House that such a strike would set back Iran's program by a few years. The Bush administration refused the request.[3]

Instead, Stuxnet, a highly sophisticated malicious computer worm allegedly developed by the United States and Israel, was reportedly launched in June 2009. Stuxnet was used specifically to target centrifuges at Iran's uranium-enrichment facility at Natanz. It manipulated valves on the centrifuges, increasing and decreasing their speed, putting additional pressure on them, and ultimately damaging the machines until they broke down.

The software was so sophisticated that until the machines failed, the operators did not see any problems. As described by the *New York Times*: "The computer program also secretly recorded what normal operations at the nuclear plant looked like, then played those readings back to plant operators, like a pre-recorded security tape in a bank heist, so that it would appear that everything was operating normally while the centrifuges were actually tearing themselves apart."[4]

As a security measure, Iran had kept the targeted computers disconnected from the internet—they were "air-gapped." Because the worm had to gain physical access to one of the machines to infect the network, Stuxnet was designed to spread via infected USB drives. "That was our holy grail," one of the architects of the plan said. "It turns out there is always an idiot around who doesn't think much about the thumb drive in their hand."[5]

Although Stuxnet crashed about a thousand centrifuges and delayed Iran's program at the Natanz plant, it did not stop Tehran's continued buildup of enriched uranium.[6] However, it had a huge impact on how we think about cyberattacks.

"Previous cyberattacks had effects limited to other computers," said Michael V. Hayden, former CIA director. "This is the first attack of a major nature in which a cyberattack was used to effect physical destruction," rather than just slow another computer, or hack into it to steal data. "Somebody crossed the Rubicon," he said.[7]

Unfortunately, and inevitably, Stuxnet did not stay at Natanz. It was released into the wild when, reportedly, an engineer at an infected facility connected his work laptop to his home network. Stuxnet is now all over the world.[8]

The achievement of the Iran nuclear deal in 2015 ended the US standoff with Tehran under the Obama administration (although the Trump administration withdrew from the deal in 2018). But if the diplomatic approach had not worked, Obama reportedly had another cyber card up his sleeve: Nitro Zeus.[9]

If hostilities seemed imminent or war broke out, Nitro Zeus was allegedly designed to disable Iran's air defenses, communications systems, and crucial parts of its power grid. "The idea was to use cyberweapons to paralyze Iran in the opening hours of any conflict, in hopes that it would obviate the need to drop any bombs or conduct a traditional attack," wrote the *New York Times*.[10] It was never used, but Nitro Zeus involved thousands of American military and intelligence personnel, cost tens of millions of dollars, and would have "placed electronic implants in Iranian computer networks to 'prepare the battlefield,'" in Pentagon-speak.[11]

Military planners warned that Nitro Zeus could have had significant effects on civilians, particularly if the United States wound up disabling large parts of Iran's electrical grid and communications networks.

Interestingly, only the president can authorize an offensive cyberattack such as this, just as the president must approve the use of nuclear weapons. But in 2018, President Trump approved a classified document known as National Security Presidential Memoranda 13, reportedly giving the US Cyber Command more leeway to conduct offensive online operations without receiving presidential approval.[12] For example, Cyber Command can now place implants of malicious software inside foreign networks without lengthy approval processes that run up to the president.[13]

"This was an enormous, and enormously complex, program," said one participant in the Nitro Zeus project. "Before it was developed, the US had never assembled a combined cyber and kinetic attack plan on this scale."

Cyber warfare has become a standard element of the arsenal for what are now called "hybrid" conflicts.[14]

In 2014, the Obama administration reportedly started a similar cyber offensive against North Korea's program to develop long-range ballistic missiles.[15] But ultimately, Pyongyang had a successful test of an ICBM, the Hwasong-15, on November 28, 2017.

The United States' use of cyberattacks against nuclear and missile programs in Iran and North Korea has changed the nature of warfare. As the United States sets the example for hybrid conflicts, and as the tools to fight them proliferate, this boomerang is sure to come back at us. No country is more dependent on computer systems, and thus more vulnerable to cyberattack, than the United States. Today, the United States is a leader in cyber-offensive capabilities. But we, and most experts, believe that it is only a matter of time before Stuxnet, or some comparable malware, is turned against the United States.

"The Department of Defense should expect cyberattacks to be part of all conflicts in the future, and should not expect competitors to play by our version of the rules, but instead apply their rules," warned a 2013 report of the Pentagon's Defense Science Board.[16] Adversaries have discovered that the United States' advantage in digital connectivity is also a weakness that can be exploited in asymmetric ways. In the same way that relatively small states or groups can sow chaos through acts of terrorism, those that cannot compete with the United States in conventional warfare will increasingly turn to cyberattacks.

"I understand the urgent threat," said Amy Zegart, a Stanford University intelligence and cybersecurity expert, speaking of America's offensive use of cyber. "But thirty years from now, we may decide it was a very, very dangerous thing to do."[17]

## CYBERATTACK ON NUCLEAR WEAPONS

Imagine if an adversary could hack into the computers that warn the United States about possible nuclear attacks. We might not see a real attack if it was coming. Or we might "see" an attack when none is there. What if hackers could take control of US nuclear weapons—to launch them, or to prevent their launch when so ordered? Cyber threats—which did not exist when the systems were first built—dramatically increase the danger of accidental nuclear war and could undermine deterrence itself.

The risk of cyberattack on US nuclear infrastructure is real. Such attacks could target early warning, communications, and delivery systems, with devastating consequences. Cyber threats increase the risk of false warning and miscalculation and magnify the danger of sole authority. Reducing these risks will require changes to US nuclear policy and posture.

Mikhail Gorbachev wrote in 2019 that "nuclear weapons could go off because of a technical failure, human error or computer error. The last alarms me the most. Computer systems are now used everywhere. And how many times have computers and electronics failed—in aviation, in industry, in various control systems?"[18]

In 2013, the Pentagon's Defense Science Board issued a major study on the Department's preparedness for cyberattacks. The results were not reassuring. The study found that weapon systems were vulnerable and the Department was "not prepared to defend against this threat."[19]

Alarmingly, the report warned that in a cyberattack, military commanders could lose "trust in the information and ability to control US systems and forces," including nuclear weapons and command, control, and communications systems. So, as a result of cyberattacks, the president could be faced with false warnings of attack or lose the ability to control nuclear weapons.

This shocking reality is still true today and will only become more dangerous as weapons systems become more interconnected. A 2018 report by the Nuclear Threat Initiative asked the key question: "If we cannot be confident that systems will work under attack from a sophisticated opponent, and if we cannot have full confidence in our ability to control nuclear weapons systems, what does this say about the continued viability of nuclear deterrence?"[20]

"Indeed, it is not inconceivable to see this as the beginning of a possible transition to a condition of mutually *un*assured destruction (MUD): a context where states may no longer feel that they will always be able to threaten nuclear retaliation to deter nuclear attack," writes Andrew Futter of the Royal United Services Institute in London.[21]

The Trump administration's 2018 Nuclear Posture Review (NPR) recognizes the cyber threat to nuclear weapons, but minimizes the risk and proposes the wrong solution. "The emergence of offensive cyber warfare capabilities has created new challenges and potential vulnerabilities for the NC3 [nuclear command, control, and communications] system. Potential adversaries are expending considerable effort to design and use cyber weapons against networked systems. While our NC3 system today remains assured and effective, we are taking steps to address challenges to network defense, authentication, data integrity, and secure, assured, and reliable information flow across a resilient NC3 network."[22]

Let's unpack Trump's NPR a bit. It says, "Potential adversaries are expending considerable effort to design and use cyber weapons against networked systems." Got it. The threat is real. Then, the NPR says, "Our NC3 system today remains assured and effective." Not so. As independent studies show, US cyber vulnerabilities are not "potential"; they are real now. And that means the current system cannot be "assured" and we cannot be confident it will be "effective" if needed. Nor is the administration taking adequate steps to address the threat.

For example, the 2013 Defense Science Board report found that "the cyber threat is serious, with potential consequences similar in some ways to the nuclear threat of the Cold War."[23] It continued, "The United States cannot be confident that our critical Information Technology systems will work under attack from a sophisticated and well-resourced opponent." The report recommended "immediate action to assess and assure national leadership that the current US nuclear deterrent is also survivable" against realistic cyber threats.[24]

Technical cybersecurity defensive measures must be undertaken, but they are not enough. The fundamental problem is that cyber defenses simply cannot keep up with cyber offenses, not the least because we never know with enough detail what the offenses will be. Cyber threats to nuclear weapons increase the risk of use due to miscalculation, increase the risk of unauthorized use, and could undermine confidence in the nuclear force.

"Nuclear weapons systems are likely to remain vulnerable to cyber threats regardless of what cybersecurity improvements are made in the future," said Herb Lin, a cyber policy scholar at Stanford University, "so much so that *changes in nuclear posture* [italics ours] are necessary to compensate for risks introduced by the cyber threat."[25]

## THOUSANDS OF ATTACKS

Right now, this very minute, adversaries are trying to disrupt US infrastructure and information networks, including nuclear weapons systems. "We see, literally, thousands if not millions of attacks against our systems every day," Gen. John Hyten, then commander of US Strategic Command said in 2019.[26]

North Korea, like many other countries, is a prolific hacker as national policy. In 2014, apparently in retaliation for a movie that mocked Mr. Kim, North Korean hackers hit Sony Pictures Entertainment, destroying Sony's computer servers, paralyzing the studio's operations, and eventually leaking embarrassing emails from executives. This became a model for the Russian attacks and leaks of emails ahead of the 2016 elections.

The "WannaCry" attack, which paralyzed more than 150 organizations around the globe in 2017, was also traced to North Korea.[27] Victor Cha of the Center for Strategic and International Studies said in a 2019 *New York Times* interview that cyberattacks have become the "third leg" of North Korea's overall military strategy. "They're never going to compete with the United States and South Korea soldier to soldier, tank for tank," he said. "So they have moved to an asymmetric strategy of nuclear weapons [first leg], ballistic missiles [second leg], and the third leg is cyber, that we really didn't become aware of until Sony."[28]

So far, there have been no publicly disclosed cyberattacks against US nuclear weapons and related command and control systems. But that does not mean they have not occurred. As detailed in chapter 3, there was a false warning of a nuclear attack in 1979 that has not been fully explained, as the government could not determine whether the cause was "human error, computer error, or a combination of both."

More recently, in 2010, missile operators lost contact with fifty Minuteman ICBMs at F. E. Warren Air Force Base in Wyoming for nearly an hour because of a computer failure. These fifty missiles carry enough firepower to kill some twenty million people. The issue affected "communication between the control center and the missiles," the Air Force said at the time.[29] The event was "a significant disruption of service," said an Air Force official, who added that "something similar happened before at other missile fields."

For almost an hour, the systems that protect against unauthorized launch of America's missiles were compromised, elevating the risk of a nuclear accident. When the Wyoming rockets went off-line, the underground launch centers that control them could no longer detect and cancel any unauthorized launch attempts.[30] The crews could not have fired the missiles, had the president so ordered, or intervened if an enemy was trying to launch them.

"Over the course of three hundred alerts—those are twenty-four-hour shifts in the capsule—I saw this happen to three or four missiles, maybe," former Air Force missile command officer John Noonan said of the communication lapse. "This is fifty ICBMs dropping off at once. I never heard of anything like it."

The Air Force determined that an improperly installed circuit card in an underground computer was responsible, and the problem was fixed. But President Obama pushed for more and found that the Minuteman silos' internet connections could have allowed hackers to put them out of commission for days or weeks. In the mid-1990s, the Pentagon uncovered a cyber vulnerability that could have allowed hackers to gain control over the key radio transmitter in Maine used to send launch orders to ballistic missile submarines. The Navy had to change its procedures so that submarine crews would not act on a launch order received out of the blue unless it could be verified independently.[31]

The risk would grow exponentially if an insider were to insert an infected thumb drive, as happened with Stuxnet in Iran, or otherwise open critical computers to outside control.

Imagine if, in a crisis, US military officers began to lose contact with half of the ICBM force, and then some submarine-launched ballistic missiles (SLBMs) on submarines. This would be a strong indication that a coordinated cyberattack against US nuclear command and control was underway, and that soon all ability to launch nuclear weapons could be lost.

While testifying before the Senate Armed Services Committee in 2013, Gen. C. Robert Kehler, then head of the US Strategic Command, expressed confidence that America's nuclear arsenal was well protected against a cyberattack, and yet he acknowledged, "We don't know what we don't know."[32] Hackers might also be targeting nuclear systems in other countries, which could be just as deadly for the United States. Kehler was asked whether Russia and China had the ability to prevent hackers from launching one of their nuclear missiles. Kehler paused for a moment and then replied, "Senator, I don't know."[33]

## DIGITAL DENIAL

Unfortunately, despite clear evidence of the need for the Pentagon to play catchup to cyber threats, the US military establishment is in dangerous denial. Since the 2013 Defense Science Board report, US weapons systems overall have become more automated and connected, and thus more vulnerable to cyberattacks. According to a 2018 report by the Government Accountability Office, GAO and others have been warning the Pentagon of cyber risks for decades. But until recently, the Department of Defense did not prioritize weapons systems cybersecurity and even now is still determining how best to address the threats.[34]

Weapons systems—including nuclear weapons—share many of the same cyber vulnerabilities as other information systems. Weapons systems are large, complex, "systems of systems" that come in a wide variety of shapes and sizes. But many weapons systems rely on commercial and open-source software and are subject to many cyber vulnerabilities. Weapons systems also rely on firewalls and other common security controls to prevent cyberattacks. Weapons system security controls can also be exploited or bypassed if the system is not properly configured. Finally, weapons systems are operated by people—a significant source of cybersecurity vulnerability for any system.

GAO reported that "mission-critical cyber vulnerabilities" had been found in many weapons systems being developed by the Pentagon. Using relatively simple techniques, "white hat" hackers were able to take control of these systems and largely operate undetected. These cyber vulnerabilities could allow hostile hackers to do the same.

GAO found that, in some cases, system operators were unable to effectively respond to the hacks. Moreover, the Department of Defense does not know the full scale of its weapons systems vulnerabilities because, for a number of reasons, tests were limited in scope and sophistication. "Cyberattacks can target any weapon subsystem that is dependent on software, potentially leading to an inability to complete military missions or even loss of life," the GAO found.

In other words, just about every weapons system under development by the Pentagon is vulnerable to cyberattack.

Unbelievably, numerous weapons systems used commercial software but did not change the default password when the software was installed. This glaring oversight allowed test teams to look up the password on the internet and gain administrator privileges for that software. "Multiple test teams reported using free, publicly available information or software downloaded from the internet to avoid or defeat weapon system security controls," reported the GAO.

Similarly, a Department of Defense Inspector General audit released in December 2018 found that the US Missile Defense Agency was failing to take basic cybersecurity precautions to protect its sensitive information. The audit found that the agency "did not protect networks and systems that process, store, and transmit (missile defense) technical information from unauthorized access and use." Administrators for classified networks had no intrusion-detection systems in place to watch for cyberattacks, much less stop them, according to the report.[35]

You might think such cyber vulnerabilities do not apply to nuclear weapons, but you would be wrong. According to media reports, officials said the programs reviewed by the GAO included two of the three major planned nuclear weapons delivery systems: the Columbia-class submarine and the replacement for the Minuteman ICBMs, known as the Ground Based Strategic Deterrent.[36]

As the US nuclear arsenal is replaced with new systems over the next twenty-five years, those systems will have greater cyber connections. "We have a number of nuclear systems that are in need of recapitalization," Werner J. A. Dahm, the chair of the Air Force Scientific Advisory Board, said in late 2016. He was referring to a new nuclear cruise missile, an ICBM, and the B-21 stealth bomber. In the future, he said, "these systems are going to be quite different from the ones that they may replace. In particular, they will be much more like all systems today, network connected."[37]

As Gen. Hyten testified about the US nuclear command and control system in February 2019, "And so the big challenge that we have is: how are we going to replace that old ancient thing that works so well that we know works, but won't work after about another decade? How do we replace that with something that works just as well and with modern technology when we have the cyber threats we have to look at? One of the great things about it being so old is the cyber threats are actually fairly minimal."[38]

Hyten continued: "So I have some ideas . . . The broad base structure of that idea is to . . . develop a number of pathways for a message to get through that is so nearly infinite that nobody can ever figure out exactly where it is or deny the ability for that message to get through. That's the way to do things in the future and I think we'll have the means to do that. I'd have to talk about in a much more classified level to get . . . into the details."

This "infinite pathways" approach to cyber vulnerabilities might work to stop an attacker from *preventing* a message from getting through. But it is also likely to provide many more entry points for an attack, and thus increase the danger of false messages. In other words, blocked launches may become less likely, but false alarms more likely.

As the 2017 Defense Science Board (DSB) report warns, "As the United States recapitalizes new nuclear capabilities, these should not be networked by default. (Connectivity may make such capabilities more modern, but it also widens the attack surface for adversaries.) The United States does not need 100 percent confidence to provide effective deterrence. Leaders would do well to focus first on minimizing adversary confidence in their ability to disrupt or deny our systems."[39]

## NUCLEAR IS NOT THE ANSWER

Although efforts are being made to reduce US cyber vulnerabilities, the United States will simply be unable to close the technical gap anytime soon. According to the 2017 DSB report, written by senior retired defense, intelligence, and military officials, "The unfortunate reality is that, for at least the next decade, the offensive cyber capabilities of our most capable adversaries are likely to far exceed the United States' ability to defend key critical infrastructures." Moreover, these adversaries may be able to thwart our military response through cyberattack.[40]

Meanwhile, said DSB, "regional powers (e.g., Iran and North Korea) have a growing potential to use indigenous or purchased cyber tools to conduct catastrophic attacks on US critical infrastructure." In addition, the report found that "a range of state and non-state actors have the capacity for persistent cyberattacks and costly cyber intrusions against the United States, which individually may be inconsequential (or be only one element of a broader campaign) but which cumulatively subject the Nation to a 'death by 1,000 hacks.'"

If the cyber danger to nuclear weapons clearly exists but cannot be addressed by cyber defenses alone, then what can we do?

The Trump administration's answer is to threaten the first use of nuclear weapons in response to a (nonnuclear) cyberattack. The NPR states that the Trump administration would consider the use of nuclear weapons in "extreme circumstances," such as "significant nonnuclear strategic attacks," which include attacks on infrastructure and "US or allied nuclear forces, their command and control, or warning and attack assessment capabilities."[41] The NPR also states that the administration will "posture our nuclear capabilities to hedge against multiple potential risks and threat developments," including cyberattacks.[42]

There are multiple problems with trying to deter cyberattack with nuclear weapons.

**Attribution.** First, figuring out who to blame for a cyberattack can be difficult and time-consuming. Attackers will try to hide their trail and even seek to place the blame on someone else. According to then STRATCOM commander Gen. John Hyten, "to effectively deter, you have to be able to see, characterize and attribute where the threat is coming from."[43]

It's one thing to have adequate confidence in the identity of an attacker to allow public shaming or even a retaliatory cyberattack; it is quite another story to have enough confidence to start a nuclear war.

**Credibility.** Second, even if you could determine the perpetrator with absolute confidence (unlikely), threatening a nuclear response would be wildly out of proportion to even a massive cyberattack and thus not likely to be credible, convincing, or effective in stopping the attack in the first place. Threatening the use of nuclear weapons against a nuclear-armed state would be suicidal. Threatening a non-nuclear state goes against decades of US policy, as it would provide a good reason for that state to get nuclear weapons.

**Targetability.** Third, if the perpetrator is a terrorist or other subnational group, we might have no idea where to find it. And even if we did, that location could be inside a sovereign state that had nothing to do with the attack.

**Response.** Fourth and finally, if the US military reaches for a weapon to retaliate against a cyberattack, that weapon might have been disabled by the attack itself. "Russia and China will also be working to increase their ability through cyberattack (and other means) to delay, disorganize, disrupt, and where possible negate US military capabilities," concluded DSB. "Such cyberattacks may target military systems specifically, or the civilian critical infrastructure on which civil and military activities depend. An attack on military systems might result in US guns, missiles, and bombs failing to fire or detonate or being directed against our own troops . . . the successful combination of these attacks could severely undermine the credibility of the US military's ability to both protect the homeland and fulfill our extended deterrence commitments."[44]

Now imagine that the lower end of cyber responses (cyberattacks, non-nuclear weapons) are disabled, and nuclear weapons are included at the high end. Things could escalate very quickly—and dangerously. "If US offensive cyber responses and US nonnuclear strategic strike capabilities are not resilient to cyberattack, the president could face an unnecessarily early decision of nuclear use—assuming that US nuclear capabilities are sufficiently resilient."[45] We must do all we can to make sure no president is faced with a nuclear-or-nothing choice. Better that nuclear not be on the menu in the first place.

For all of these reasons, we cannot depend on cyber defense, cyber deterrence, or even cyber response, and we cannot include nuclear weapons as part of that response. We must turn instead to adapting our nuclear posture to cyberattacks that we cannot stop.

## CYBER IS THE "NEW WILD CARD IN THE DECK"

Governments must operate under the assumption that digitized systems, including nuclear weapons, are vulnerable, will be hacked, and may already be compromised.

The dangers posed by current US nuclear policies are bad enough, and when the possibility of cyberattack is added in, things get downright ugly. Cyber is the new threat multiplier that should serve to motivate changes to US nuclear plans.

As former STRATCOM commander Gen. James Cartwright wrote in 2015, "In some respects the situation was better during the Cold War than it is today. Vulnerability to cyberattack, for example, is a new wild card in the deck."[46]

Above all else, the new cyber reality should force us to question the wisdom of making any decisions quickly in the nuclear realm. As the Nuclear

Threat Initiative puts it, "No plausible improvements in the cybersecurity of these systems will allow leaders to ignore the possibility of the cyber threat. Increasing decision time (including potential changes in alert status) may be the only way to compensate for risks introduced by the cyber threat."[47]

In a world with cyberattacks, things are not as they seem. Incoming missiles on computer screens may not really be there. Rather than a false alarm based on a computer glitch, it could be a sophisticated cyberattack timed to coincide with an international crisis. Backup information from satellites and radars could also be corrupted. The military may lose contact with some or all of the nuclear arsenal.

We must work to buy the president—any president—as much decision time as possible. Sole authority is dangerous because it allows a snap decision based on little or no consultation. Weapons on alert and launch on warning drive the president to make a choice too soon. In the world of cyberattacks, we should assume nuclear alerts are false until proven otherwise. This is the only way to avoid blundering into nuclear catastrophe.

Weapons on alert, poised for prompt launch, do not deter cyberattackers, who can hide behind layers of false identities. But these same weapons offer a tempting target for cyberattack and can play havoc with US global security.

If we took an enlightened approach, we would reset our entire nuclear posture. Instead of quick launch, we would establish a retaliation-only posture. We would not launch nuclear weapons first, but only in response to a verified attack.

This simple guiding principle—a second-strike deterrent posture; shoot second, never first—should be used to drive a more rational and safer process to rebuild the US nuclear arsenal, as we will see in chapter 7.

But at a more basic level, in light of cyber threats to nuclear weapons and to government control over these weapons, we must question the continued viability of nuclear deterrence itself. If the president cannot be confident that the nuclear button, once pushed, will work, then what good is it? The cyber revolution has the potential to make nuclear deterrence obsolete.

# PART II

# A NEW NUCLEAR POLICY

# CHAPTER 5

# NO FIRST USE

*Here's the guarantee that we will never use nuclear weapons*
*first. This is my family, my wife, children, and grandchildren.*
*I don't want them to die. No one on earth wants that.*
—President George H. W. Bush, as told by Andrei Sakharov[1]

In a September 2016 presidential campaign debate, moderator Lester Holt of NBC News noted that President Obama "reportedly considered changing the nation's longstanding policy on first use" of nuclear weapons but instead left open the option for the United States to be the first nation to go nuclear, as it did at Hiroshima seventy-five years ago. Then he asked candidate Trump: "Do you support the current policy?"

"I would certainly not do first strike," Mr. Trump declared. "I think that once the nuclear alternative happens, it's over," he said, at first appearing to agree with Obama, and US public opinion generally, that the United States should not use nuclear weapons first. But then Trump followed the well-worn path of his predecessors, saying, "At the same time, we have to be prepared. I can't take anything off the table."[2]

That left Trump in almost the same position as Obama and all presidents before them. They would never use nuclear weapons first, but at the same time they would not give up the option to use them first, or to threaten to do so.

In 1989, Soviet scientist Andrei Sakharov met President George H. W. Bush, who showed him a photograph of the extended Bush family. Bush said: "Here's the guarantee that we will never use nuclear weapons first. This is my family, my wife, children, and grandchildren. I don't want them to die. No one on earth wants that." Sakharov replied, "But if you insist that you will not strike first, you must make an official announcement of that, put it into the law." Bush had no reply.[3]

Would any rational president actually use nuclear weapons first? We think not. Since Truman used atomic bombs in 1945, US presidents—who must ultimately own the decision—have shown almost universal revulsion at the thought of using them again. Every time since Nagasaki and Hiroshima that US presidents considered using atomic weapons—in the Korean War, against China in 1958, over Cuba in 1962, in Vietnam—they found them unnecessary or unusable. This stands in stark contrast to nuclear practitioners (STRATCOM commanders, strategic planners, etc.), who have by and large sought to make nuclear weapons more "usable" and thus more relevant to US military plans.

This is understandable. Presidents comprehend the destructive power of the bomb and they cannot imagine ordering a full-scale attack, particularly against Russia, and inviting a massive retaliation. Some presidents have sought "limited" nuclear options to make nuclear threats more believable. Yet, since Truman, presidents have refused to use the bomb even in small numbers or against nonnuclear adversaries—even in the face of military defeat.

So, ultimately, if no sane president would ever order the first use of the bomb, then why does US policy allow first use? It should not.

"Successive presidents of both parties have contemplated and then categorically rejected the employment of nuclear weapons even in the face of grave provocation," wrote Gen. Butler. "In today's security environment, threats of their employment have been fully exposed as neither credible nor of any military utility."[4]

During my time as secretary of defense and since, I (Bill) never confronted a situation, nor could even imagine a situation, in which I would recommend that the president make a first strike with nuclear weapons—understanding that such an action, whatever the provocation, could bring about the end of our civilization. When Libya was discovered to be secretly building a chemical weapons facility, I said its leaders would be wise to stop, because we would not allow them to make it operational. When reporters asked me whether we were prepared to use nuclear weapons to stop them, my answer was simple and to the point: "We would not need to. Our conventional forces are more than adequate to the task."

As Representative Adam Smith (D-WA), chairman of the House Armed Services Committee, said in an interview with us, "I don't think there is any rational reason why we would use nuclear weapons first."[5]

## THE PARADOX OF FIRST USE

One of the great contradictions of US atomic history is that presidents have refused to give up the option to start nuclear war even though they have no need or intention to exercise that option. The United States has had a de facto no-first-use policy for decades that no president has been willing to formalize.

President Truman, as we have seen, was loath to use the bomb a third time and yet adopted an official policy of first use that we still have today. The 1948 Berlin crisis made it clear that the Soviet Union was the dominant conventional military power in Europe, that Moscow might use that power even though it did not yet have the bomb, and that the United States could not stop Moscow with conventional force alone. The United States sought to deter a feared Soviet conventional invasion of Europe with nuclear threats.[6]

This was the beginning of the Cold War nuclear arms race. The Soviets tested their first nuclear device in 1949. Then the United States increased its production of nuclear bombs and long-range bombers and started the development of thermonuclear weapons, the H-bomb. Washington strategists thought (wrongly) that if the United States could maintain nuclear superiority it could also maintain a credible threat to use nuclear weapons first.

In 1950, the Truman administration rejected a proposal by advisor George Kennan (author of the Cold War doctrine of containment) to adopt a no-first-use policy and continued to state that the use of nuclear weapons was under "active consideration."[7] Even so, President Truman ended his time in the White House with these words: "The war of the future would be one in which man could extinguish millions of lives at one blow, demolish the great cities of the world, wipe out the cultural achievements of the past—and destroy the very structure of civilization that has been slowly and painfully built up through hundreds of generations. Such a war is not a possible policy of rational men."[8]

In 1953, the Eisenhower administration decided to produce and deploy in Europe large numbers of tactical atomic bombs, including nuclear land mines, artillery shells, and rockets, for use on the battlefield. The administration also declared a policy of "massive retaliation" in which it committed to respond to a Soviet attack with immediate and massive nuclear force. This became known as "security on the cheap," because atomic bombs were less expensive than the troops and tanks that would be needed to counter Soviet conventional forces. In an effort to maintain a credible threat against the Soviets, and to reassure European allies, the US deployed several thousand tactical nuclear weapons to Europe by the late 1960s. In response, the Soviets deployed their own large arsenal of tactical weapons. If war had broken out, Europe would likely have been destroyed by the very weapons sent to protect it.

And yet, after the Soviets tested a thermonuclear device on August 12, 1953, President Kennedy's national security advisor McGeorge Bundy wrote, "To Eisenhower, general nuclear war will soon be no better than suicide . . ."[9] Eisenhower said in a 1954 National Security Council meeting that "we would never enter the war except in retaliation against a heavy Soviet atomic attack."[10] His hawkish secretary of state John Foster Dulles agreed, saying, "our main purpose" in having nuclear weapons was "to deter the use of that weapon by our potential enemies."[11]

President Eisenhower, like other presidents of the atomic age, did not want nuclear catastrophe to define his legacy. As he said before the United

Nations in 1953, "Surely no sane member of the human race could discover victory in such desolation. Could anyone wish his name to be coupled by history with such human degradation and destruction?"[12] And as Eisenhower said in 1957, "You just can't have this kind of war. There aren't enough bulldozers to scrape the bodies off the streets."[13]

The master plan for carrying out nuclear war, the Single Integrated Operations Plan, or SIOP, was first developed in December 1960 at the end of the Eisenhower administration. The scale of the plan was immense. The United States could launch its entire force of more than three thousand warheads against the Soviet Union, China, and their allies. Eisenhower sent his science advisor, George Kistiakowsky, to Strategic Air Command (SAC) in Offutt, Nebraska, in November 1960 to see how the plan was going. Kistiakowsky came back with the impression that the plan would "lead to unnecessary and undesirable overkill."[14] President Eisenhower confided to an aide that the plan and its implications "frighten the devil out of me."[15]

Meanwhile, global opposition to the bomb was growing. In 1961, the United Nations General Assembly passed a resolution stating that the use of nuclear weapons was "contrary to the spirit, letter and aims of the United Nations and, as such, a direct violation of the Charter." It stated that such use would be indiscriminate in its effects and thus is "contrary to the rules of international law and the laws of humanity" and that any state using nuclear weapons would be "committing a crime against humanity and civilization." Japan voted in favor of this resolution but would later oppose US moves in this direction.[16]

## PRESIDENTS AND THE BOMB

Of course, we will never know whether a US president would actually have ordered the use of nuclear weapons in Europe if the Soviets had launched a conventional invasion. Without such an order, the Soviets would likely have taken much of western Europe. But the use of US nuclear weapons to stop this would have likely led to global nuclear war and total devastation of the United States and Russia, and much of the rest of the world.

McGeorge Bundy retells a story in which President Kennedy asked former secretary of state Dean Acheson, during the 1961 Berlin crisis, when the US might need to use nuclear weapons. Acheson, considering his words carefully, said he believed "the president should himself give that question the most careful and private consideration, well before the time when the choice might present itself, that he should reach his own clear conclusion in advance as to what he would do, and that he should tell no one what his conclusion was. The President thanked him for his advice, and the exchange ended."[17]

This discussion shows just how personal and solitary any presidential decision to use nuclear weapons would be.

Starting with President Kennedy, all incoming presidents have received a top-secret briefing on the US nuclear arsenal and how to use it. After President Kennedy was briefed on the war plan in 1961, he commented, "And we call ourselves the human race."[18] Kennedy and his defense secretary Robert McNamara tried to revise the war plan to provide options short of all-out nuclear attack. Kennedy wanted the option of targeting Soviet weapons and not cities, called "counterforce," that he thought was more humane. But targeting all Soviet conventional weapons would require a huge buildup of nuclear forces and start an expensive new arms race. McNamara preferred what he called "assured destruction," which required weapons to destroy 20 to 25 percent of the Soviet population and 50 percent of industry.[19] "Mutual" got tacked on and "mutual assured destruction," or MAD, was born. This sense of mutual vulnerability became for many the defining feature of the Cold War.

J. Robert Oppenheimer, the technical director of the Manhattan Project, which produced the first atomic bombs, once compared the United States and Russia to "scorpions in a bottle, each capable of killing the other, but only at the risk of his own life." Oppenheimer noted that scorpions, a species known for belligerence, would sooner or later take that risk.[20]

Mutual vulnerability meant that if one side went nuclear, the other would too, and no one could win. Thus, nuclear war must never start. McNamara recalled that "in long private conversations with successive Presidents—Kennedy and Johnson—I recommended, without qualification, that they never initiate, under any circumstances, the use of nuclear weapons. I believe they accepted my recommendations."[21]

In 1962, Kennedy stated publicly that "there is not going to be any winner of the next war. No one who is a rational man can possibly desire to see hostilities break out, particularly between the major powers which are equipped with nuclear weapons. . . .

"Now, if someone thinks we should have a nuclear war in order to win, I can inform them that there will not be winners in the next nuclear war, if there is one, and this country and other countries will suffer very heavy blows." Kennedy continued, "So we have to proceed with responsibility and with care in an age where the human race can obliterate itself."[22]

For McNamara, nuclear weapons were simply not usable: "Nuclear warheads are not military weapons in the traditional sense and therefore serve no military purpose other than to deter one's opponent from their use."[23]

President Nixon got his first briefing on the war plan in January 1969. "It didn't fill him with enthusiasm," said Henry Kissinger, Nixon's secretary of state. A National Security Council (NSC) staffer recalled that Nixon was "appalled" by the nuclear briefing because he realized that his only available options were for massive nuclear strikes involving thousands of weapons. A few weeks later, at an NSC meeting when the discussion turned to nuclear war scenarios, Nixon declared: "No matter what [the Soviets] do, they lose their cities. . . . What a decision to make."[24]

Kissinger felt the same way. As he explained during an August 1971 meeting, "I have often asked myself under what circumstances I would go to the President and recommend that he implement the SIOP, knowing that this would result in a minimum casualty level of 50 million."[25]

Discussing Nixon's thinking several years later, Kissinger, referring to the SIOP, said, "If that's all there is [Nixon] won't do it." Soon after the briefing he called former secretary of defense McNamara and asked, "Is this the best they can do?" A few years later, Kissinger made a rare reference to ethical concerns by declaring that "to have the only option of killing 80 million people is the height of immorality."[26]

On May 11, 1969, Nixon flew on the National Emergency Airborne Command Post (NEACAP), a converted Boeing 707, and took part in a nuclear war simulation. According to his chief of staff, H. R. Haldeman, it was "pretty scary. They went through the whole intelligence and operational briefings—with interruptions, etc. to make it realistic." Haldeman wrote that Nixon "asked a lot of questions about our nuclear capability and kill results. Obviously worries about the lightly tossed-about millions of deaths."[27]

At the same time, Nixon wanted to be the master of the nuclear threat. It was a key aspect of his "Madman Theory." As Haldeman wrote:

> Nixon not only *wanted* to end the Vietnam War, he was absolutely convinced he *would* end it in his first year. . . . The threat was the key, and Nixon coined a phrase for his theory. . . . We were walking along a foggy beach after a long day of speechwriting [during Nixon's presidential campaign in 1968]. He said, "I call it the Madman Theory, Bob. I want the North Vietnamese to believe I've reached the point where I might do *anything* to stop the war. We'll just slip the word to them that, for God's sake, you know Nixon is obsessed about Communism. We can't restrain him when he's angry—and he has his hand on the nuclear button—and Ho Chi Minh himself will be in Paris in two days begging for peace."[28]

It is worth noting that this is not the way the Vietnam War ended. So much for the Madman Theory. Nixon found that using the bomb was not feasible—and thus his threats were not credible—as it would have entailed great political costs in domestic and international public opinion. Kissinger cautioned Nixon against such actions that would create large civilian casualties because, he said, "I don't want the world to be mobilized against you as a butcher."[29]

Alexander Haig, a hawkish advisor to Nixon who later would serve as Reagan's secretary of state, understood the contradiction of first use. He wrote that "the mere existence of our superior power often bailed us out of potential disaster even though we were determined, in the depths of the national soul, never to use it."[30] Haig believed that no American president would use the bomb except to defend Europe from a Soviet attack.

Then this theory was put to the test. In the 1970s, the Soviet Union deployed a new nuclear weapon: the SS-20 intermediate-range missile. Fielded in large quantities in the western part of the Soviet Union, the SS-20 could strike every European NATO country, but did not directly threaten the United States. To our NATO allies, this raised a troubling question: If the Soviet Union struck European NATO countries but not the United States, would Washington really use nuclear weapons against the Soviet Union and thereby risk a Soviet retaliation? Or to put it another way: Would an American president risk New York, Washington, or Chicago to save London, Paris, or Hamburg?

To put this question to rest, the United States, with NATO approval, decided to respond to the SS-20s by deploying US intermediate-range ground-launched cruise missiles (GLCMs) and Pershing-II ballistic missiles in Europe. These forces could attack Moscow and other targets inside Russia, Ukraine, and Belarus. This was intended to "couple" the United States more closely to Europe, by making sure that any nuclear conflict would include the Soviet Union, and therefore, the United States.

But it was a high-risk strategy, with the Soviet Union and NATO facing off with hundreds of missiles poised to bring death and destruction to all of Europe, including Russia.

In summer 1980, President Carter was looking for ways to limit a potential nuclear war. He signed Presidential Directive 59 in July, which revised the range of targeting choices a president would have. Carter's new plan focused on attacking the Soviet political leadership and envisioned limited nuclear strikes.

President Reagan was especially hostile to the notion that he might have to make nuclear use decisions in a crisis. Indeed, before he became president, top advisors wondered whether he would order nuclear retaliation in response to warning of a Soviet missile strike on the United States. It is possible that SIOP briefings increased his anti-nuclear sentiments.

Reagan's strong support for missile defenses (see chapter 8) was driven in part by his deep antipathy for nuclear weapons. Before he became president, he reflected that "we have spent all that money and have all that equipment, and there is nothing we can do to prevent a nuclear missile from hitting us." He understood the choices a president would face in a nuclear war: "The only options he would have would be to press the button or do nothing. They're both bad."[31]

Reagan rejected the concept of MAD. He recoiled at the idea that, as president, he would have to make a fate-of-the-world decision about nuclear weapons in a crisis. "I swear I believe Armageddon is near," he wrote in his diary in 1981 when Israel bombed an Iraqi nuclear reactor. As president, Reagan carried no wallet, "no keys in my pocket—only secret codes that were capable of bringing about the annihilation of much of the world as we knew it," he wrote.

In March 1982, Reagan was part of a nuclear war simulation called Ivy League. It did not go well. "In less than an hour President Reagan had seen the United States of America disappear," recalled White House advisor Thomas C. Reed. "I have no doubt that on that Monday in March, Ronald Reagan came to understand exactly what a Soviet nuclear attack on the US would look like."[32]

After the exercise, Reagan got a briefing on the war plan, "which was as scary as the earlier presentation on the Soviet attack," said Reed. "It made clear to Reagan that with but a nod of his head all the glories of imperial Russia, all the hopes and dreams of the peasants in Ukraine, and all the pioneering settlements in Kazakhstan would vanish. Tens of millions of women and children who had done nothing to harm American citizens would be burned to a crisp."[33]

After watching the ABC television drama *The Day After* in 1983, Reagan got another briefing on the war plan. Reagan recalled that "in several ways, the sequence of events described in the briefings paralleled those in the ABC movie. Yet there were still some people at the Pentagon who claimed a nuclear war was 'winnable.' I thought they were crazy. Worse, it appeared there were also Soviet generals who thought in terms of winning a nuclear war."[34] Two years later, Reagan and Gorbachev would issue their historic statement that "nuclear war cannot be won and must never be fought."

Reagan had to carry the nuclear launch identification codes (the "biscuit") until his last day in office. He was eager to get rid of the authentication card and thereby relieve himself of this burden. But his national security advisor, Lt. Gen. Colin Powell, told him: "You can't get rid of it yet." Reagan had to wait until George H. W. Bush was sworn into the presidency before he could hand it off to a military aide.[35]

In the waning years of the Cold War, Gorbachev and Reagan, seeing how dangerous these weapons were to the world, discussed abolishing all nuclear weapons in their summit at Reykjavik. They did not succeed in achieving that lofty goal, but those discussions led directly to the 1987 Intermediate-Range Nuclear Forces (INF) Treaty, which eliminated all intermediate-range missiles. So, the SS-20s, Pershings, and GLCMs were all dismantled, ending (or so we thought) that dangerous period in Europe.

## TURNING THE TABLES

The INF Treaty was followed by more dramatic actions. The Warsaw Pact and the Soviet Union broke apart in 1992, signaling the end of the Cold War. Almost overnight, the conventional military balance shifted in favor of the West. Today, there is no longer a conventional threat imbalance in Europe to correct with the first use of US nuclear forces. Now we have the reverse. Alexander Haig's scenario for first use no longer exists.

The end of the Cold War meant that presidents could rethink the wisdom of first use. Les Aspin, the first defense secretary to serve the first post–Cold War president, Bill Clinton, conducted a nuclear posture review (NPR) and said that a no-first-use policy could be part of a new nonproliferation policy. However, neither that NPR nor the two that followed included a no-first-use declaration, in large part because US allies expressed concerns that this might undermine their security. That misguided thinking is still a barrier to a no-first-use policy today.

Unfortunately, in 1993, a weakened Russia reversed its no-first-use policy to counter the now stronger conventional forces in western Europe. Russia's security situation continued to get worse as NATO expanded to Poland, Hungary, the Czech Republic, and the Baltic states. Russia has said that it would use the bomb first only if the existence of the Russian state were at risk due to a conventional conflict that it was losing. This is similar to NATO's policy of first use during the Cold War.

## OBAMA AND NO FIRST USE

After a top-secret nuclear policy briefing in 2008, Obama told a close advisor that it was perhaps one of the most sobering experiences of his life. He said, "I'm inheriting a world that could blow up any minute in half a dozen ways, and I will have some powerful but limited and perhaps even dubious tools to keep it from happening."[36]

President Obama announced in April 2009, his first year in office, his intention "to seek the peace and security of a world without nuclear weapons." So when his administration began a review of US nuclear policy, there were hopes of major changes, including a no-first-use declaration. However, when the 2010 NPR was released, it fell short of the mark. The review did state that the United States would not use the bomb against nations that did not have nuclear weapons and were complying with their nonproliferation obligations (such a policy has existed in various forms, and with various caveats, since 1978). It also said that there was no need to use nuclear weapons to counter a conventional attack by a nuclear state.

But Obama's NPR did not endorse a no-first-use policy. The final draft of the posture review reportedly contained this view, but in the end one of Obama's senior advisors on the National Security Council talked the president into preserving the first-use option. This advisor reportedly made the argument that a quick US nuclear strike might be needed to stop a terrorist operation from spreading biological pathogens.

The possible use of nuclear weapons against chemical or biological weapons (as opposed to conventional weapons) has been a persistent barrier to a no-first-use declaration and a justification for the US policy of "strategic ambiguity." For example, then secretary of state James Baker told Iraq's foreign minister just before the 1991 US–Iraq war that if "you use chemical or biological weapons against US forces, the American people will demand vengeance and we have the means to exact it." Baker said, "It is entirely possible and even likely, in my opinion, that Iraq did not use its chemical weapons against our forces because of that warning. Of course, that warning was broad enough to include the use of all types of weapons that America possessed."[37]

Baker may believe that his nuclear threat worked in this case, but there is no way to know. As for US intentions, the memoirs of senior officials confirm that Bush had decided by December 1990, before the war started, that he would not use nuclear or chemical weapons even if Iraq used chemical weapons.[38]

Instead of declaring a no-first-use policy, or that the "sole purpose" of nuclear weapons is to deter their use by others (that is, for retaliation only), President Obama decided that US nonnuclear capabilities were not yet up to the task. In other words, there were still some scenarios where Obama's advisors wanted to preserve the option of using nuclear weapons first, such as to counter the threat of biological weapons. However, it seems to us that threatening to use nuclear weapons in response to a biological (or chemical) attack from a nonnuclear adversary is not credible (as no rational president would do it) and not necessary (we could use conventional weapons instead) and thus would have little military value. Moreover, this highly unlikely scenario prevented President Obama from reaping the benefits of declaring a no-first-use policy, such as reducing the more pressing threat of accidental war and building support for US nonproliferation policies.

Despite rejecting a no-first-use declaration, the NPR made clear that the Obama administration would seek to create the conditions for a sole-purpose policy, as it would serve US national security interests.

In 2016, the last year of his administration, President Obama tried again on a no-first-use declaration.[39] Former defense officials with full information about US military capabilities and the most likely threats, such as Bill and former STRATCOM commander Gen. James Cartwright, were in full support of no first use. According to Gen. Cartwright, "Nuclear weapons today no longer serve any purpose beyond deterring the first use of such weapons by our adversaries."[40]

But it was not enough. Obama's personal support for no first use was overwhelmed by caution from his own cabinet members (defense, state, and energy) and from allies, particularly Japan. Japanese prime minister Abe reportedly communicated his opposition personally to Obama because he feared no first use could increase the chances of conventional conflict with China or North Korea. Moreover, Japanese opposition was reportedly based on the concern that no first use would somehow weaken the US commitment to Japan's defense.[41]

Jon Wolfsthal, who was part of Obama's internal deliberations, told us that the Obama administration did not reject no first use but, rather, ran out of time to make an informed decision. "By the time we worked that up to the President's level, I had informed [national security advisor] Susan Rice and the President by memo that I did not believe that we had the time remaining in the Obama administration to make a decision that we wanted to achieve no first use and actually get allied support. That was going to take time."[42]

Despite the mixed views from his own cabinet, Obama could still have adopted a no-first-use policy but did not. Instead, Vice President Joe Biden gave a speech in January 2017, just before Trump assumed office. Biden said, "Given our nonnuclear capabilities and the nature of today's threats—it's hard to envision a plausible scenario in which the first use of nuclear weapons by the United States would be necessary. Or make sense. President Obama and I are confident we can deter—and defend ourselves and our Allies against—nonnuclear threats through other means."[43]

To Wolfsthal, this statement meant that "they'd lived through some crazy things over eight years. They understood the world as well as anybody who has ever occupied the White House and they understood the allies' concerns, they understood the threats coming to the US, and they didn't believe that nuclear first use was relevant to handling or addressing those threats."[44]

Ultimately, Obama wound up in the same place as his predecessors: he had no intention of ever using nuclear weapons first but could not bring himself to spend the political capital necessary to make it official policy. The Obama administration never made a decision on no first use, one way or the other.

## TRUMP AND NO FIRST USE

As a candidate, Donald Trump refused to rule out the first use of nuclear weapons, yet he seemed disinclined to use them: "I will do everything in my power never to be in a position where we will have to use nuclear power. It's very important to me."[45] After winning the White House, he implied his willingness to initiate the use of nuclear weapons against North Korea when he threatened that country with "fire and fury like the world has never seen."

The Trump administration's 2018 NPR made dangerous changes to US nuclear policy, such as increasing the number of situations in which Washington might use nuclear weapons first. The Trump NPR envisions using nuclear weapons first against nuclear weapons states in response to nonnuclear attacks, such as cyberattacks, as well as against nonnuclear states. This is a dramatic and unwelcome change from previous administrations.

The Trump administration has deployed new "low-yield" nuclear weapons to counter what it perceives to be a chance that Russia might use a similar nuclear weapon to stop a conventional war that it was losing. However, Moscow has denied that it would escalate a conflict to the nuclear level to end or "deescalate" a crisis.

According to Russian ambassador Anatoly Antonov,

Another flagrant example of unwillingness to hear us is the notion that our nuclear doctrine includes an "escalate to deescalate" concept that includes the possibility of a first "limited low-yield nuclear strike." This belief is held widely enough to be mentioned in the US Nuclear Posture Review released in February 2018. Clearly, this allegation does not withstand criticism. All those who doubt it could have a look at Article 27 of the Russian Military Doctrine. It plainly states that our country "reserves the right to use nuclear weapons in response to a use of nuclear or other weapons of mass destruction against it and (or) its allies, and in case of an aggression against it with conventional weapons that would put in danger the very existence of the state." Therefore, as Vladimir Putin said, "our strategy does not include a preemptive use of nuclear weapons. . . . Our concept is a retaliatory counterstrike.[46]

Despite Moscow's denials, Trump's proposal for low-yield bombs brought back an old, discredited argument against no first use: that nuclear war, if it starts small, could be controlled and might not spin out of control into all-out nuclear holocaust.

## THE MYTH OF LIMITED NUCLEAR WAR

Advocates of limited nuclear war take advantage of the happy fact that there has never been a nuclear exchange between two atomic states, and thus we do not really know how it would go. They suggest that if one side used a "low-yield" nuclear weapon, the other side would stay at that level because they would not want to escalate to full-scale war for the obvious reason that it would mean utter destruction for both sides.

But to us, this is exactly why no one should ever use a nuclear weapon—of any yield—in the first place. There is every reason to believe that, once attacked with atomic weapons, a nation would be so outraged and/or would assume a full attack was on the way that it would respond with everything they've got. Expecting a limited response is wishful thinking and dangerous in the extreme.

Some policy analysts, sitting at their desks, write about "escalation control," but the reality is that, in a real war, these analysts would not be making the decisions. Indeed, it is unlikely that the president would have full control over all decisions on weapons use. In the heat and fog of war, decisions could be made by the commanders doing the fighting, dealing with the tactical situation as it evolves and using the weapons they have to their best advantage.

Fundamentally, it is unlikely that there is such a thing as a limited nuclear war, and preparing for one is folly. "A nuclear weapon is a nuclear weapon," said George Shultz, who served as President Ronald Reagan's top diplomat. "You use a small one, then you go to a bigger one. I think nuclear weapons are nuclear weapons and we need to draw the line there."[47] Former defense secretary Jim Mattis said, "I don't think there's any such thing as a tactical nuclear weapon. Any nuclear weapon used at any time is a strategic game changer."[48]

As Congressman Adam Smith said in a nuclear policy conference keynote in 2019, "One of the things that I would like to communicate to the Russians is, yeah, it doesn't work that way. Use a nuclear weapon, we are going to respond and we are not going to be overly concerned about being 'proportional.' A nuclear weapon is a nuclear weapon, and if you use it, we will nuke you. So don't."[49]

For decades, senior US policy makers have recognized the insanity of preparing for limited nuclear war, and yet the idea will not die. As McGeorge Bundy, George Kennan, Robert McNamara, and Gerard Smith wrote in a joint *Foreign Affairs* article in 1982:

> It is time to recognize that no one has ever succeeded in advancing any persuasive reason to believe that any use of nuclear weapons, even on the smallest scale, could reliably be expected to remain limited. Every serious analysis and every military exercise, for over twenty-five years, has demonstrated that even the most restrained battlefield use would be enormously destructive to civilian life and property. There is no way for anyone to have any confidence that such a nuclear action will not lead to further and more devastating exchanges. Any use of nuclear weapons in Europe, by the Alliance or against it, carries with it a high and inescapable risk of escalation into the general nuclear war which would bring ruin to all and victory to none.
>
> The one clearly definable firebreak against the worldwide disaster of general nuclear war is the one that stands between all other kinds of conflict and any use whatsoever of nuclear weapons. To keep that firebreak wide and strong is in the deepest interest of all mankind. In retrospect, indeed, it is remarkable that this country has not responded to this reality more quickly. Given the appalling consequences of even the most limited use of nuclear weapons and the total impossibility for both sides of any guarantee against unlimited escalation, there must be the gravest doubt about the wisdom of a policy which asserts the effectiveness of any first use of nuclear weapons by either side. So it seems timely to consider the possibilities, the requirements, the difficulties, and the advantages of a policy of no first use.[50]

## TIME FOR NO FIRST USE

There is no need to use US nuclear weapons first in any realistic scenario. The United States is the world's dominant conventional military power and does not need nuclear weapons to deter or respond to nonnuclear threats to itself or its allies, including threats from biological weapons. Moreover, by leaving open the first-use option against nonnuclear states, the Trump administration undermines US nonproliferation goals. How can the United States possibly convince other nations that they do not need nuclear weapons if the United States itself says it needs them for nonnuclear threats?

Some will argue that the United States, even if it never uses nuclear weapons first, has profited from making nuclear threats in the past and should not give up the option in the future. The United States has threatened first use in the Berlin blockade, the Korean War, the Vietnam War, the Cuban Missile Crisis, the Iraq War, and others.[51] Few Americans have not heard the phrase often repeated by presidents that, in a military crisis, "all options are on the table."

Clearly, there is a strong reluctance among US presidents to pull the nuclear trigger, as evidenced by the fact that the bomb has not been used in an attack since 1945 and was not used in the face of military stalemates, like Korea, or defeats, like Vietnam. But just as clearly, there has been no similar restraint against nuclear threats.

Since 1945, fortunately, none of these numerous threats have been carried out. Many were surely bluffs, but maybe not all. Were some successful? It's hard to know. In some cases, the threats may not have changed the planned behavior of the adversary. Even if there was a change in behavior, it may have been for unrelated reasons. Yet it is possible that in a few cases the threats were effective, and either way, they were seen as effective by senior administration officials.

The perceived need to make nuclear threats, and the belief that they are essential to US national security and to that of our allies, explains in part why no president has so far been willing to make a formal no-first-use pledge. The United States has resisted such a commitment despite its clear benefits in convincing other states not to pursue nuclear weapons and in building support for the Nuclear Nonproliferation Treaty (NPT). Lack of leadership by the United States in this area makes it very difficult for the United States to lead on the wider goals of delegitimizing the bomb and stopping its spread.

Today, the declining value of nuclear threats makes them no longer worth the price. It is time for the United States to realize that the national security costs of maintaining a first-use option far outweigh any benefits. Given the well-justified reluctance of US presidents to use the bomb, US threats to use it first are simply not credible and should not be believed. By renouncing first use, Washington would gain clear benefits, such as making it harder for a delusional president to launch a unilateral nuclear attack; increasing our leverage on Russia to do the same, thereby reducing the risk of accidental war; and enhancing our ability within the international community to support nonproliferation efforts.

As Senator Dianne Feinstein said, "The only moral purpose for nuclear weapons is to deter their use by other nuclear powers. To unambiguously state that we will never initiate nuclear war would reinforce an international norm that will help keep us safe and minimize the very real risk that a foreign power might someday misinterpret a benign rocket launch or a nonnuclear military action as a nuclear strike."[52]

US allies, such as Japan, who oppose a US no-first-use policy, need to realize that US nuclear threats that are not credible do not make them safer. In fact, incredible nuclear threats make other threats less credible. Allies would be better served by supporting no first use and working for stronger nonproliferation policies. Yes, the United States should come to Japan's (and other allies') defense in a crisis, but as long as the crisis does not involve a nuclear attack on Japan, there is no need for US nuclear threats.

As the only nation to have suffered a nuclear attack, and as a supporter of the elimination of nuclear weapons, Japan should support no first use as a step toward that goal. If all nations declared a no-first-use policy, and those declarations were credible, then nations would not need their arsenals and could work together to eliminate them. In opposing no first use, Japan is opposing the principle of nuclear disarmament itself.

Finally, by endorsing a no-first-use policy, the United States would be honoring a strong American value: that the United States should never start a nuclear war. About two-thirds of Americans say that the Unites States should make a clear declaration that it will never use the bomb first.[53]

# CHAPTER 6

# HOW *NOT* TO SPEND $2 TRILLION

*The rationale for the Triad, for the number of nuclear weapons we have, that rationale doesn't exist anymore.*
—Representative Adam Smith, Chair, House Armed Services Committee[1]

What would you do with $2 trillion? You might buy the major sports teams. ALL of them. Or you might buy Apple Computer AND Amazon. If you are concerned about the environment, you might buy an electric car—or 50 million electric cars. You could pay off all US student loans or fund America's health care or rebuild our crumbling infrastructure. Or you could buy 150 aircraft carriers.

The United States plans to spend about $2 trillion over the next three decades to maintain and replace its nuclear arsenal.[2] This is more than the country would be spending on its entire State Department during that period. Even so, if we thought this level of spending were required to ensure US national security, we would support it. Instead, we believe that rebuilding the Cold War nuclear arsenal as presently planned would actually *decrease* our security.

How could this be? Current plans call for building new nuclear weapons as if the Cold War never ended. This plan not only entails huge costs, but it also entails great dangers. Our Cold War nuclear arsenal provided unambiguous deterrence, but it came with a serious security risk: that a nuclear war could start by miscalculation or accident. And that risk is not merely theoretical. The Cold War brought us to the brink of nuclear war at least five times. In the past, we lived with these great risks because some thought they were necessary. But we should not continue to take these risks. It is time to take a serious look at the nuclear arsenal, replace only the weapons we need for today's threats, and retire the rest.

We support a strong but restrained US nuclear deterrent as long as nuclear weapons are held by other nations. But we do not support replacing every weapon in the arsenal. A dollar spent on nuclear weapons is likely a dollar taken away from other military needs, such as sustaining conventional forces and countering terrorism and cyberattacks.

During the Cold War, many believed that the greatest danger to the United States was a surprise nuclear attack from the Soviet Union. Our nuclear forces were designed to deter Moscow from launching a first strike. In retrospect, it is clear that the Soviet Union would *not* have launched an unprovoked attack then and will not now. It has no incentive to do so. Today, the greatest danger is not a Russian surprise attack but a US or Russian blunder—that we might accidentally stumble into nuclear war. As we make decisions about which weapons to buy, we should use this simple rule: If a nuclear weapon increases the risk of *accidental* war and is not needed to deter an *intentional* attack, we should not buy it.

The nation now has the historic opportunity to learn the right lessons from the Cold War and build the right arsenal. Sadly, current plans are doubling down on the wrong lessons. We are still acting—and planning to spend trillions of dollars—as if the greatest danger is a Russian surprise attack, not an accidental nuclear war. But we can't plan for both of these things at the same time. We need to choose. By planning for a surprise attack, we are making an accidental war more likely. We are still playing radioactive Russian roulette.

There is still time to get this right, if we act soon. The United States has embarked on a dangerously misguided nuclear modernization plan. How would this plan change if we shifted the primary threat to accidental war?

## THE PROBLEM WITH ICBMS

We call it the Upper Midwest. In nuke-speak, it's known as the "nuclear sponge."

The United States currently deploys hundreds of nuclear missiles in silos across Colorado, Montana, Nebraska, North Dakota, and Wyoming. Each missile carries a nuclear payload many times more powerful than the Hiroshima bomb, capable of killing millions of people. The Pentagon is now planning to build a new, deadlier generation of these missiles, which are to be housed in these silos.

Unlike submarines and bombers, which are mobile, these ICBMs are sitting ducks. Russia knows exactly where they are and can attack them at any time, but is deterred from doing so because US subs and bombers on alert would survive to retaliate against such a suicidal assault. But what would a US president do if he or she believed that such an attack was underway—that is, if he or she received an alarm from our missile warning system?

In response to warning of an incoming attack, the president has two main options: (1) launch the ICBMs before the presumed attack arrives (known as launch on warning or launch under attack), or (2) wait to make sure it is a real attack. If the president waits and the attack is real, most of the ICBMs would be destroyed.

The choice should be easy. Launching nuclear weapons on warning of attack is simply too risky because our early warning systems are vulnerable to false alarms and cyberattacks, not to mention human error. Jim Miller, a senior Pentagon nuclear policy official in the Obama administration, remarked in March 2019, "The Nuclear Posture Review in which I was involved attempted to move us away from launch under attack and did that in small ways. I think it's important to continue to make that effort."[3]

A 2013 report from the Pentagon reflected this shift, stating, "Recognizing the significantly diminished possibility of a disarming surprise nuclear attack, the guidance directs DoD to examine further options to reduce the role of Launch Under Attack plays [sic] in US planning, while retaining the ability to Launch Under Attack if directed."[4]

One of the main ways that the Obama administration moved away from launch on warning was to ensure that it could meet its requirements for retaliation without using the ICBMs. In other words, the United States can fully respond to a nuclear first strike even if all of its ICBMs are destroyed. Thus, there is no need for a president to feel pressured into launching ICBMs in the face of an incoming attack. The ICBMs are simply not needed for an effective response.

"We no longer need to use land-based ICBMs to achieve our mission as defined by the president," Jon Wolfsthal said in our interview. "We can deter an attack with survivable systems, meaning submarines. . . .

We were able to take the ICBMs out of the frontline must-have/must-fire forces in order to achieve damage limitation, which allowed you to basically say, 'We can ride out the attack,'" Wolfsthal continued. "And one of the nice things about that, although we didn't do this, was you could basically say, 'Oh well, then ICBMs can go away. You don't need them.'"[5]

Moreover, launching the US ICBMs would not stop the incoming attack, if there really is one. It would simply be one component of the response, which would include nuclear weapons launched from bombers and nuclear subs.

Former secretary of defense Jim Mattis, who supports keeping the option of launching on warning, told us, "We would likely never need to launch on warning. If the missiles are in the air, then deterrence has failed and our retaliation is not going to stop those in the air from striking us. So [we would need to] take a deep breath and decide what we're going to do. . . .

"But for the deterrent to work," Mattis continued, "we must make it very clear that we will end life in your capital city and other places if you ever tried. An enemy attack does not condemn us to an immediate act or emotional reaction; it should be a studied, deliberate process. An immediate reactive launch would be necessary only if needed to stop additional launches."[6]

However, an immediate retaliatory strike would be highly unlikely to stop additional launches from Russia. If Moscow did not launch everything at once, it would presumably launch its vulnerable weapons first, and save its survivable weapons for later. A US retaliation fired at Russian ICBM silos would land on empty holes.

So although it may be counterintuitive, if the president is faced with a "use them or lose them" choice on ICBMs, the best decision by far is to lose them. If it's a false alarm, that restraint will have saved the world. If it's a real attack, we don't need the ICBMs anyway, so better to let them get buried by radioactive rubble.

But why, then, do we have ICBMs at all?

According to official policy, even if these missiles are never launched, they still serve a useful purpose—to be destroyed in the ground, along with the missileers and all the people that live anywhere near them. Their purpose is to "absorb" a nuclear attack from Russia, acting as a giant "nuclear sponge" or a "missile sink."[7] As STRATCOM commander Gen. Hyten said in March 2019: "That's one of the big values of our ballistic missiles. Four hundred ballistic missiles create a huge targeting problem for any adversary. The only way to get after four hundred hardened nuclear missiles is with a whole bunch of incoming weapons. And if you decide to attack those, then you pretty much are guaranteeing that we'll attack back. That's deterrence in a nutshell and that creates a huge element of our deterrent process . . . It would be a missile sink."[8]

When asked at an October 2019 press conference why states would want to host ICBMs and thereby put themselves at greater risk, Representative Smith said, only partly in jest, "Apparently, they want to be targeted in a nuclear first strike."[9]

Can it possibly make sense to draw a nuclear attack *toward* the United States, rather than away from it? Even during the Cold War, analysts challenged this plan, claiming it was "madness to use United States real estate as 'a great sponge to absorb' Soviet nuclear weapons."[10]

Yet the nuclear sponge is still with us. Not only that, but the Trump administration is planning to spend up to $150 billion to do it all over again.

In 2016, before he became defense secretary, Jim Mattis asked the Senate Armed Services Committee if it was time to remove the land-based missiles, as "this would reduce the false alarm danger."[11] After he became secretary, unfortunately, he backed away from that view.

Instead, Mattis defended the new ICBM and the nuclear sponge mission, although he did not call it that. Testifying before the Senate on January 12, 2017, Mattis said, "It's clear they are so buried out in the central United States that any enemy that wants to take us on is going to have to commit two, three, four weapons to make sure they take each one out. In other words, the ICBM force provides a cost-imposing strategy on an adversary."[12]

Cost-imposing indeed, but for whom? Yes, attacking US ICBMs would be very costly for Russia, mainly because the United States would retaliate with hundreds of nuclear weapons launched from submarines at sea. But what about the costs to Colorado, Montana, Nebraska, North Dakota, and Wyoming?

The United States does not need ICBMs and can safely phase out the existing missiles without replacing them. This would save $150 billion; it would take the missile states out of the crosshairs; and it would remove the danger that our ICBMs could trigger an accidental nuclear war.

How would President Trump, or any president, respond if he or she were woken from a deep sleep at 3 a.m. and told that our warning system indicated that hundreds of Russian nuclear missiles were landing in minutes? Would the president understand that it could be a false alarm; would the president know that if he or she launched the missiles, they could not be called back or aborted; would the president know that these weapons were superfluous anyway? Or would he or she reflexively launch a counterattack? No one else has the authority to make this call, and once the missiles are launched, a massive nuclear war is sure to follow.

We are concerned that senior officials in the nuclear weapons business do not take the risk of false alarms seriously enough. They reassure us that the

chance of a false alarm is "at an all-time low" and that "the statistical probability that the United States would launch ICBMs as a result of a false alarm
is close to zero."[13] Such language is dangerous.

The truth is that the probability *is* very low, but it is *not ze*ro. And the
consequences of a false launch would be astronomical. Human errors and
machine errors in our warning system have occurred and will occur again. It
is only a matter of time before the odds add up to a catastrophic failure. But
here is the point: we do not have to take that terrible risk anymore.

The United States is rebuilding its nuclear-armed submarines that can
hide under the oceans, able to survive a Russian nuclear attack. That is all we
need to keep Moscow in check. In the unlikely event that new threats emerge
that could put the subs at risk, the Air Force is rebuilding the insurance policy: nuclear-capable bombers. The ICBMs are at best extra insurance that we
do not need; at worst, they are a nuclear catastrophe waiting to happen.

## SECOND STRIKE ONLY

With no ICBMs, no launch on warning, and a policy of no first use, the
United States would shift to the much safer position of launching second,
never first. The sole purpose of US nuclear weapons would be to deter their
use by others—as most Americans assume is already the case. All of our forces
would be geared to be survivable such that an adversary could never be confident of its ability to disable US nuclear forces in a first strike.

As Gen. Hyten said in 2019, "As long as we have nuclear capabilities that
our adversaries cannot attack, they cannot take out and they cannot eliminate, we'll be able to prevent the use of nuclear weapons on our nation."[14]

He continued:

I remember when I interviewed for this job with President Obama and
then I interviewed with Secretary Mattis after he took over, they asked
me: what is the number one reason . . . we have nuclear weapons?

And I said the reason we have nuclear weapons is to prevent people
from using nuclear weapons on us. That's exactly why we have them. If

you don't have a robust capability and our adversaries don't believe that you're willing to respond, then you run the risk that somebody will take that step across the line that nobody ever wants to experience.

We have been operating within this dynamic for decades. Back in 1982, Bundy, Kennan, McNamara, and Smith wrote, "The Soviet government is already aware of the awful risk inherent in any use of these weapons, and there is no current or prospective Soviet 'superiority' that would tempt any-one in Moscow toward nuclear adventurism. (All four of us are wholly unper-suaded by the argument advanced in recent years that the Soviet Union could ever rationally expect to gain from such a wild effort as a massive first strike on land-based American strategic missiles.)"[15]

This is still true today. Jim Miller said in March 2019, "The scenario where we're on the precipice of a nuclear exchange is extremely unlikely. One of the key reasons why it is so unlikely is that each side has a secure second-strike capability. Until someone has a better model, we ought to rein-force that."[16]

Moscow agrees. Russian ambassador Antonov said in 2019 that "MAD, mutually assured destruction, is still alive. And taking into account this con-cept, you'll see that I can't see any possibility for anybody to attack neither United States nor Russia."[17]

But what, exactly, constitutes a second strike? Some would say that a US launch on warning of attack is retaliation. But if the warning is false, then the "retaliation" would actually be a first strike. Thus, US policy should prohibit any nuclear launch until a nuclear attack on the US has been confirmed. This may sound horrific, but it is consistent with past thinking. As Secre-tary of State Dean Rusk recalled, "I always believed we are committed to a second strike, actually to absorb nuclear destruction on our territory before counterattacking."[18]

As long as other nations have nuclear weapons, the United States must have a nuclear capability such that no one thinks they can attack us and get away with it. But the extent of our nuclear arsenal and means of launching these world-destroying weapons are past due for change.

## DO WE NEED THE "TRIAD"?

Shifting to a second-strike retaliatory force means we can make major changes to US nuclear forces and policy, such as moving away from the "triad." The United States has a triad of nuclear-armed delivery systems, including ICBMs, submarines, and bombers. Bombers appeared first and remain the only means by which nuclear bombs have ever been used in war. Next, land-based ballistic missiles (ICBMs) appeared and, for the Navy, ballistic missiles launched from submarines.

The nuclear bureaucracy has developed elaborate justifications for why each leg of the triad is essential. But those rationales were developed after the fact. The triad emerged over time mainly as the result of interservice rivalry between the Air Force and the Navy in the 1950s and 1960s. As nuclear weapons became a central arena of Cold War competition, defense spending began to flow, and no branch of the military wanted to be left out. Even the Army had nuclear bombs for decades (including the infamous nuclear backpack and atomic artillery). Eventually, the Navy and Air Force stopped competing for nuclear dollars and joined forces, arguing that all three legs of the stool were essential. Since then, the triad has become so entrenched it is rarely questioned and is taken as an article of faith. As the CATO Institute aptly describes it, "'Maintaining the triad' is uttered as an almost religious incantation to affirm one's commitment to the very idea of nuclear deterrence."[19]

It was not until the end of the Cold War that some people inside government started to question the triad. In the Clinton administration, Defense Secretary Les Aspin initiated the first Nuclear Posture Review (NPR), which briefly considered alternatives to the triad.

But the military services and key members of Congress were hostile to scaling back nuclear delivery systems. Senator Strom Thurmond (R-SC) prompted US Strategic Command chief Adm. Henry Chiles to testify during a committee hearing that "ICBMs are necessary in our force for the future."[20] Other senators sent letters supporting the triad to President Clinton. In the end, the Clinton administration was not willing to fight a major battle with Congress on the issue.

Each leg of the triad has a corresponding constituency made up of powerful members of Congress, senior military leaders and Pentagon officials, defense industry lobbyists, and think tank experts. The US government spends at least $50 billion per year on nuclear weapons and related expenses, which means a lot of people have a strong interest in seeing that funding continue. From this perspective, it is no accident that, after the Soviet Union ceased to exist, the triad outlived the enemy it was designed to counter.

To put this in context, the amount of money spent by organizations seeking to limit nuclear weapons pales in comparison. Independent foundations, like Ploughshares Fund, provide funding to nongovernmental organizations, like the Nuclear Threat Initiative, the Arms Control Association, and many others, to support their work for saner nuclear policies. Those grants add up to tens of millions of dollars per year. We like to think this is a sizable amount, but it is dwarfed by the government and defense industry funds (tens of BILLIONS per year) spent on building, promoting, marketing, and lobbying for nuclear weapons.

## LOOK BEFORE YOU LEAP

The next president should review current US plans to rebuild the nuclear arsenal, looking for ways to reduce nuclear dangers and save money. If this examination leads to a reduction in presently planned nuclear programs and costs, it would be consistent with the 2016 Democratic Party platform, which stated that the party "will work to reduce excessive spending on nuclear weapons–related programs that are projected to cost $1 trillion over the next thirty years."[21] That figure did not include recent cost increases or the significant cost of inflation over three decades.

In addition, the chair of the House Armed Services Committee, Representative Adam Smith (D-WA), had this to say in March 2019 about the Trump administration's nuclear plans: "This is an area where I think we can save money and still meet our national security objectives."[22]

The United States is in the very early stages of a program to build and maintain a new generation of missiles, submarines, and bombers, which will cost about $50 billion a year for the next several decades. This effort accelerated soon after the United States ratified the New START Treaty in 2010 (see chapter 7). To secure votes for New START, President Obama made a political commitment to modernize or replace the US arsenal. These programs were started in the Obama administration, but, according to officials, Obama did not necessarily support all of these new weapons.

Wolfsthal said that toward the end of Obama's second term, in 2016, the President was worried that his team had "taken our eye off the ball here" and that the plan had grown too big. Obama wanted "to dig deep, make sure there's no gold plating here, understand what we need and what our options are."[23]

Wolfsthal and his colleagues spent six months putting the Pentagon, the Office of Management and Budget (OMB), and the National Nuclear Security Administration (NNSA) "through the ringer and I am still not popular with most of those people. But the reality is that the President wanted to understand what the options were, and I think we did a really good job laying out what you actually get for each dollar and what each leg of the triad actually provides, and you can do an assessment based on what we developed. In fact, those documents are still available to President [Trump]."[24]

After President Obama was briefed on his options, his advisors were split; some thought the full rebuild program was justified, others did not. Obama said, according to Wolfsthal, "Well, look, we're not going to make these decisions now, we're months away from handing over a government and what we need to do is make sure these options are teed up for our successor so that they understand that they're not locked into one particular program or action."[25] So, they kicked the can down the road to the next president, whom they expected to be Hillary Clinton. Instead, it was Donald Trump.

Today, these decisions are being considered by the Trump administration, which was not involved in the eight years of debate in the Obama administration. And in 2021, either the Trump administration or a new administration will have to make these important decisions. Our advice to a new administration would be to not overinvest in nuclear weapons systems and thus encourage a new arms race: the United States should build only the levels needed for second-strike deterrence. We should encourage Russia to do the same, but even if it does not, US levels of nuclear forces should be determined by what we *need*, not by a misguided desire to match Moscow missile for missile. If Russia decides to build more than it needs, it is *their* economy that will suffer, just as it did during the Cold War.

## RUSSIA'S WEAPONS TODAY

Russia has also begun building a new generation of missiles, submarines, bombers, bombs, and warheads, for both strategic and tactical nuclear forces. The Russian state media has embarked on an aggressive program to advertise, and even flaunt, these new weapons. This program was established during a period when the Russian economy was booming, based on very high revenues from oil and natural gas, which are primary contributors to the Russian federal budget. But if oil prices fall again, Russia may have to reconsider its ambitious and costly nuclear upgrade.

This extensive rebuilding program clearly has been influenced by the significant deterioration of relations between the United States and Russia. This deterioration was a direct result of Russia's annexation of Crimea, incursions into eastern Ukraine, and threats to the Baltic nations, but it has been heavily influenced by longer-standing disputes over NATO expansion, European ballistic missile defense deployment, and American support for the so-called color revolutions.

The tense relations today is a *causative* factor in the present nuclear arms buildup, but they also make it *more dangerous*. We do not believe that a nuclear war would be started deliberately by either Russia or the United States, but it is all too conceivable that a nuclear war could be started *accidentally* or through *miscalculation*, and today's tense relations create the conditions that could lead to a dangerous political miscalculation.

Our present nuclear arsenal was conceived and built during the Cold War, and we should not assume that it is the right arsenal for today's needs. There have been fundamental changes in technology and in geopolitics these past four decades. One fundamental change is the strength of NATO conventional forces: during the Cold War, NATO conventional forces were only a third the size of Warsaw Pact forces, and, in the early years, not qualitatively better. Today, NATO has significantly stronger conventional forces than Russia does, both quantitatively and qualitatively.

Another fundamental change is that during the Cold War, we were faced with Warsaw Pact forces as well as those of the Soviet Union. Today, most of the Warsaw Pact nations (such as Poland, Romania, the Czech Republic, and Hungary) and many of the former Soviet Republics (like Ukraine) are not allied with Russia. And certainly the West today has a commanding lead over Russia in economic strength and technological innovation. What remains the same is that Russia has a strategic nuclear arsenal essentially equal to that of the United States, and a tactical nuclear force significantly larger.

The United States does not need to rebuild its nuclear forces to match those it had during the Cold War. And yet it must do what is necessary to maintain a sufficient deterrent. The question is where to draw the line.

## MAINTAINING A SECOND-STRIKE FORCE

To maintain a robust second-strike force and save hundreds of billions of dollars, the United States can take the following steps.

### LAND-BASED MISSILES

First, as explained above, the US can phase out its land-based ICBM force. During the Cold War, the United States leaned heavily on ICBMs because they provided accuracy not then achievable by sub-launched missiles and bombers, and they provided another insurance policy in case the sub force somehow became disabled. Today, we have quite high accuracy in both our submarine and bomber forces, and we have enough confidence in them that we do not need an additional insurance policy. We do not need a "belt and suspenders" for our "belt and suspenders."

Some will argue that the way to address the vulnerability of ICBMs is to give them mobility, and indeed Russia bases many of its ICBMs on trucks and trains. But as I (Bill) can attest, past attempts to find publicly acceptable basing options for mobile missiles failed dramatically (I and others made vigorous but futile attempts to build a mobile ICBM, called the MX). If any future secretary of defense thinks that seeking to place mobile ICBMs in anyone's backyard would be a good use of their time, they should check with me first.

## SEA-BASED MISSILES

Second, the US arsenal plan calls for new nuclear-armed submarines, which we support, assuming a critical analysis of the number of subs needed. No adversary could believe that a surprise attack would destroy all of the at-sea submarines. And any one of them (carrying as many as 192 thermonuclear warheads) is capable of inflicting unacceptable damage on an adversary. Moreover, because the submarines at sea are not vulnerable to a first strike, there is no reason to launch their missiles under warning of attack. This avoids serious concerns about accidental war that are inherent to silo-based ICBMs.

The submarine force alone is sufficient for assured deterrence and will be so for the foreseeable future. But as technology advances, we have to recognize the possibility of new threats to submarines, especially cyberattacks and detection by swarms of drones. The new submarine program should put a special emphasis on improvements to deal with these potential threats, assuring the survivability of the force for decades to come.

Current plans call for 12 new submarines at a cost of more than $130 billion. Each new sub will be able to carry up to 128 nuclear warheads on 16 missiles. Under the 2010 New START Treaty, the Navy plans to deploy about 1,000 warheads at sea. We find this excessive; the United States can deter nuclear attack with fewer subs and save money.

In 2011, the OMB recommended that the number of new submarines be reduced to ten.[26] The Navy pushed back by claiming that ten submarines would not be enough to support five submarines "on station" at all times. Submarines on station are deployed far off the US coasts, ready to launch their missiles on a moment's notice.

But the need for on-station submarines is mainly driven by the military's existing requirement to deploy submarine-based nuclear weapons within range of their targets so they can be launched promptly, within an hour or so. The need for twelve subs, then, has as much to do with where the warheads are deployed and how promptly they could reach their targets as it does with the number of warheads. For example, an eight-submarine fleet can carry a thousand warheads, but it can't support five subs that are forward deployed near Russia and China, ready for quick launch.

However, new nuclear guidance could relax those requirements, which are based on nuclear policy and targeting assumptions that have changed little since the Cold War ended thirty years ago. Instead of forward-deploying subs ready for prompt launch, some of them could be held back as an assured retaliatory force if ever needed.

If the next administration determines—as it should—that the United States does not need to hold so many targets in Russia and China at risk with a "prompt" submarine attack, then the requirement for twelve subs can be reduced. And if a future administration were to change its New START deployment plan, or achieve additional arsenal reductions, the requirement for deploying as many as a thousand sea-based warheads could be reduced as well.

A fleet of ten new nuclear-armed submarines will be more than adequate to meet our country's deterrent needs. The firepower on board just five or six survivable submarines would be enough to destroy the vital elements of state control, power, and wealth in Russia, China, and North Korea. In fact, just *one* boat can carry enough nuclear weapons to place two warheads on each of Russia's fifty largest cities.

The Trump administration has deployed new "low-yield" nuclear warheads on Trident missiles. These dangerous weapons are a misguided way of addressing a nonexistent problem. The United States can deter the unlikely Russian use of its low-yield bombs with its current arsenal. There are no "gaps" in the US deterrent force, and there can be no doubt in Russia's mind that the United States is serious about maintaining an unambiguously strong nuclear deterrent.

Perhaps the biggest fallacy in the argument for more "usable" weapons is the mistaken belief that a "small" nuclear war would somehow stay small; that if Russia used a "low-yield" nuclear weapon, the US would respond in kind, and that things could stay at that level. There is, of course, no experience to support this dubious theory.[27]

If the United States deploys a new low-yield nuclear weapon to counter Russia's low-yield weapons, this sends the message to Russia that nuclear war can be limited. This is a very dangerous game and not worth the risk of starting a full-scale nuclear catastrophe, as leaders are starting to see. "I would like to kill the low-yield nuclear weapon program," Representative Smith said in March 2019.[28]

Moreover, the United States already has low-yield nukes. As part of its massive arsenal, the United States has about a thousand weapons capable of being detonated at low yields, including gravity bombs and cruise missiles, both of which are being modernized at great expense.[29] If the president really ever needed to use a low-yield nuke, there are plenty.

Finally, firing a single "low-yield" warhead from a strategic submarine could undermine the survivability of the submarine itself. As Madelyn Creedon, former principal deputy administrator at the National Nuclear Security Administration, notes:

> The sea leg of the nuclear triad is the most survivable leg in large part due to the ability of Ohio-class submarines to be invisible in the open ocean. Launching a high-value D5 missile [the only type of nuclear-armed missile these subs carry] from a ballistic missile submarine will most likely give away its location. China and Russia are expanding their ability to detect a missile launch and will be able to locate a US submarine if it launches a D5 missile. Is having a low-yield warhead worth the risk of exposing the location of a ballistic missile submarine at sea?[30]

## AIR-BASED WEAPONS

Third, Trump's program calls for development of a new bomber, the B-21, with improved stealth capability. We support that program (again, assuming a critical analysis of the numbers) because it provides backup should the submarines ever suffer a temporary problem that raises a question about their capability. This is not likely, but the bomber force is an insurance policy for that contingency. The new bomber would be dual capable, usable for conventional or nuclear missions, and would provide a critical new capability for conventional forces, even if it were not necessary for deterrence. Bombers could be sent into the air in a crisis, and, once there, could loiter for many hours, allowing them to wait out an alarm while airborne. If the alarm turns out to be false, the bombers would be recalled.

There is still an open question as to whether the new bomber should be manned or unmanned. We believe that technically either is viable, providing that unmanned means *remotely controlled*. It is vital that any bomber with a nuclear mission have continuous human control, including a recall capability. That could be achieved with a remotely controlled system but not with a fully automatic system.

## COMMAND AND CONTROL

As long as nuclear weapons exist, they must remain under strict national control, even in a nuclear attack. This is essential for deterrence and to prevent unauthorized or accidental use.

Moving from first-use to second-use assured retaliation means that the command and control system can shift from quick-launch options to providing more decision time for the president. We no longer need weapons on high alert and the ability to launch on warning of attack. But we do need a survivable system that protects the president and his or her ability to issue orders under the most stressing conditions imaginable.

For the United States to shift to a second-strike-only posture, the government must have confidence that the president (or another decision maker) and the US command and control system can withstand a nuclear attack, however unlikely that attack might be. The priority must be to remove the president and his or her immediate successors from danger to a safe place where the situation can be monitored and decisions made. Depending on the attack, the president would have just minutes to escape from the White House and board a helicopter to fly to a secure location.

Some independent analysts who have seen the system close up do not believe that, once a nuclear attack has landed, the US command system would be able to respond. Bruce Blair says that today the nuclear command, control, communication, and intelligence (C3I) network "would likely collapse within a few hours of nuclear conflict. Fixing this is essential to supporting assured retaliation and enabling the president to intelligently choose a response if deterrence should fail. Instead of modernizing overkill, increasing presidential decision time should be our top priority."[31]

The United States should prioritize command and control modernization over rebuilding its nuclear weapons. The president should not feel rushed into a launch decision, and we should seek to extend the time frame well beyond an attack. This would better allow the president to reassess the post-attack situation and prudently direct the operations of surviving forces.

Improving US command and control would involve addressing a variety of emerging threats such as cyberattacks, anti-satellite warfare, and attempts to degrade or decapitate today's system. US satellites form the critical backbone of the early warning system against missile attack and are used to send messages to execute war plans. Russia, China, and now India have the ability to disrupt and destroy US satellites, and the United States can do the same to theirs.[32]

However, effective deterrence does not require that the president have absolute confidence in the command and control network. The president and his or her advisors need some degree of assurance that the system, including senior decision makers, will survive. But the key standard is that Russia must believe that the risk of launching an attack, in terms of the assured counterattack, outweighs the benefits of the attack itself. The goal is not actually to survive the attack (which ultimately may not be possible beyond a few weeks), but to prevent the attack in the first place.

A major issue for command and control is whether the president could maintain contact with submarines at sea after a nuclear attack on the United States. The primary communications link used to order the submarines to launch is vulnerable to direct attack and signal jamming. Submarine crews could not receive launch authorizations if these communication links failed, but they would still be able to fire their missiles by opening (with a blowtorch or drill) the onboard safe and retrieving the launch key.[33] This is a significant factor for deterrence. Not even a "decapitating" first strike would prevent submarines at sea from retaliating with hundreds of high-yield warheads.

Deterrence holds today even under an implausible worst-case massive surprise attack from Russia.

Frank von Hippel of Princeton University makes this important point: "Failure to communicate is not the same as failure to retaliate. What do Russian leaders think US nuclear submarine crews are going to do if they learn that the United States has been destroyed? Go to Tahiti and retire?"[34]

Using Russian leaders as our guide is a tricky business, as it is hard to judge Russian attitudes and intentions. But it is useful to recall that Russia is generally more worried about a US first strike than the other way around. So, the US command and control system does not have to be perfect, but it does have to be credible. The system today is far from perfect, and we can and should make it better. But there is no reason to believe it is not credible to Russia and China.

"By all indications, Russia understands that under any plausible scenario of nuclear war with the United States, that it would suffer unacceptable damage," said Blair. "I think everyone understands that fully. And the Russians I talked to have said that 270 US nuclear weapons delivered to targets in Russia would completely decapitate their ability to function as a society, as an economy, and as a military."[35]

The goal of overhauling the command system should be deterrence through assured retaliation, not any ability to "control" escalation. The price tag will be large and could require an increase in current annual spending that is likely to be independent of the size of the arsenal.[36] However, it must be made clear that such funding would only be justified in the context of a shift to a second-strike policy. There is little reason to invest additional billions in survivable command and control if the primary plan is to launch nuclear weapons first in a crisis.

## LESS IS MORE

As the United States plans for the future of the nuclear arsenal, we can move to a smaller but more secure second-strike force whose sole purpose is to deter attack. We do not need to spend hundreds of billions more in a dangerous and futile attempt to "prevail" in a nuclear conflict.

By phasing out ICBMs, building only ten new submarines, building fewer new bombers, and getting by with fewer weapons, we estimate that the United States could save at least $300 billion ($10 billion per year) and still field a formidable deterrent force. Savings of this size would nearly equal the combined 2019 budgets of the Department of Veterans Affairs ($199 billion), Department of State ($37.8 billion), and Department of Homeland Security ($47.5 billion).[37]

In 2007, I (Bill) joined my colleagues George Shultz, Henry Kissinger, and Sam Nunn in an op-ed alerting the world to the present dangers of nuclear weapons and calling for actions to decrease those dangers and ultimately eliminate them. For several years, the world took timely and important actions in that direction, most importantly, the Nuclear Security Summit meetings. But the sharp downturn in relations with Russia and the aggressive rebuilding of the Russian nuclear arsenal have stopped that progress. I believe we should give high priority to diplomatic initiatives that can regain earlier momentum.

We should rebuild our nuclear arsenal in a way that does not *aggravate* the present dangers, that does not *burden* us with unnecessary costs, and that keeps the door open to a return to *reductions* in nuclear arms and in nuclear dangers. Indeed, a significant success in diplomacy could allow both the United States and Russia to reconsider the kind of nuclear arsenals needed for security, and jointly scale back the new programs while they are still in their early stages.

Certain new nuclear weapons, such as the ICBM, carry higher risks of accidental war that, fortunately, we no longer need to bear. We are safer without these weapons, and it would be a mistake to deploy them.

# CHAPTER 7

# WELCOME TO THE NEW ARMS RACE

*Trump's withdrawal from the INF Treaty is dangerous and irresponsible. We should strengthen these treaties, not end them. We should invest in our children, our seniors, and our working families, not in an incredibly expensive arms race.*
—Senator Kamala Harris[1]

I f you were born after 1990, you did not live during the Cold War. You did not experience a nuclear arms race with the Soviet Union. You did not "duck and cover" under your school desk in preparation for a Soviet nuclear attack. For the most part, you have not been living under the dark cloud of possible nuclear war like your parents and grandparents did. To be clear, the dark cloud of nuclear war never really went away, but after the Soviet Union collapsed, it receded to the fringes of public concern.

If you missed out on the first nuclear arms race and feel like you've been passed over, you're in luck. It's back.

In February 2019, the Trump administration announced the United States would withdraw from the Intermediate-Range Nuclear Forces (INF) Treaty, a landmark agreement signed in 1987 by President Reagan and Soviet leader Gorbachev. Trump said, "Russia has not, unfortunately, honored the agreement, so we're going to terminate the agreement. We're going to pull out."[2]

This historic pact was arguably the most important arms treaty ever signed by the United States and the Soviet Union. It marked the beginning of the end of the Cold War, removing and eliminating an entire category of nuclear and conventional missiles from installations across Europe. And it marked the end of an incredibly dangerous period in the arms race, when all of Europe was threatened with nuclear annihilation by the deployment of thousands of nuclear weapons in Europe.

It also kick-started a new era of trust and cooperation between the two Cold War antagonists. The treaty was based on the maxim often quoted by Reagan: "Trust but verify." And indeed, this treaty entailed highly intrusive verification provisions. For example, the United States was allowed to base a contingent of on-site inspectors at the Russian missile factory at Votkinsk. These inspectors had extensive access to this once-secret missile facility, and the Russians had corresponding access to American missile factories. INF ushered in an era of transparency previously unimagined.

By withdrawing from the INF Treaty, it is not hyperbole to say that President Trump is ushering in a new Cold War. There have been other elements of US and Russian diplomacy that have moved us in this dangerous direction, but it seems to us that withdrawing from the INF Treaty was the most decisive.

Right after Trump's announcement to withdraw from INF, Russian president Vladimir Putin said his country would follow suit: "Our answer will be symmetrical," Putin said. "Our American partners declared that they will suspend their participation in the treaty, so we will suspend ours as well. They said they would start research and development, and we will do the same."[3] Of course, Putin was pleased to see Trump withdraw from INF, because Russia had already started building missiles prohibited by INF.

This is how an arms race starts.

We have seen this movie before, and it does not end well. In "Arms Race One," the United States built an insanely large nuclear arsenal that peaked at more than 30,000 warheads in the 1960s. The Soviet Union had 40,000.

How could the arms race possibly justify such huge arsenals? This is a question I (Bill) am often asked by students in my classes. I explain that, at first, massive arsenals were motivated by a misguided quest for "superiority." Both sides wanted the upper hand. When that proved unattainable, the goal became a quest for "parity." The conventional wisdom was that if we had fewer than they did, they might perceive an advantage and launch an attack. But why parity at 30,000 warheads instead of 300? The simplest answer is bad information and worst-case planning. If you don't really know what the other side has, it seems safer to assume the worst and build more. If both sides do this, you get an out-of-control arms race that at the time appeared rational to those closest to it. In retrospect, given the risk and expense entailed, it all seems crazy.

This is where arms control comes in. Arsenal limits backed by good information are the cure for a dangerous arms race. If both sides are confident that they have rough parity, then they can stop building more. And once we establish parity, we can move together to lower levels. Parity at 300 is much better than at 30,000. A smaller arsenal is cheaper to maintain, easier to keep track of, less threatening to the other side, less polluting to the environment, and would show the rest of the world that the United States and Russia are working to disarm, which supports efforts to stop the spread of the bomb to other nations.

To be clear, parity is more a political need than a strategic necessity. It has always been hard for politicians to defend a position of numerical inferiority. But we strongly believe that as long as both sides have survivable forces, parity is not essential for deterrence. As the Obama administration stated, "The need for numerical parity . . . is no longer as compelling as it was during the Cold War."[4] But what is absolutely clear is that the world is safer with parity at 300 instead of 30,000.

How much did the nuclear arms race cost? For the United States, roughly $10 trillion since 1940.[5] To understand how much money this is, imagine a single stack of 10 trillion $1 bills, laid one over the other—it would stretch for over 678,000 *miles*. You could stack them all the way to the moon (238,900 miles away) and back, and you'd still have trillions left over.

If you divide $10 trillion into equal portions for each man, woman, and child in the United States, everyone would receive over $30,000.

How did we survive the arms race? As the Cuban Missile Crisis shows us, by sheer good luck.

But global sanity got a leg up when the stars aligned and two enlightened leaders agreed to change course. In 1985, President Reagan and Soviet premier Gorbachev declared that "nuclear war cannot be won and must never be fought." This was a dramatic conceptual shift that took the wind out of the sails of the arms race. If neither side can win this race, then why are we running?

The INF Treaty followed two years later. The Cold War began to thaw and finally melted away with the breakup of the Soviet Union in 1992. Today, thanks to arms control agreements (that are now on the chopping block), both sides are down to about 6,000 warheads each; still way too many, but a huge improvement over 30,000 to 40,000. In fact, Washington and Moscow, which together possess over 90 percent of the nuclear weapons in the world, could maintain viable nuclear deterrents with just 100 nuclear weapons on each side (we should never lose sight of the utter devastation that 100 modern nuclear weapons could inflict on our two countries and the planet). Ultimately, we should seek to maintain world security without nuclear weapons, but until we achieve the political accommodations that would make that possible, deterrence with 100 nuclear weapons is a realistic goal.

The Trump administration does not seem to understand what got us this far: nuclear arms reduction agreements with Russia, negotiated mainly by Republican presidents. President Nixon signed the Anti-Ballistic Missile (ABM) Treaty in 1972, President Reagan signed the INF Treaty in 1987, and President George H. W. Bush signed the first START Treaty in 1991. Presidents Bill Clinton, George W. Bush, and Barack Obama continued the process, most recently with the 2010 New START Treaty, which brings US and Russian arsenals down to levels not seen since the 1950s.

President Trump would be the first US leader since Eisenhower to not contribute to this long, proud tradition.

These agreements provide the "rules of the road" so arsenals can be reduced in a safe, stable, and predictable way. No one wants surprises when it comes to nuclear weapons. They set equal limits on weapons, and allow for inspections so both sides can trust, and verify, the process.

## KILLING ARMS CONTROL

But now, all of this is at risk. In "Arms Race Two: The Sequel," we have the US and Russian presidents conspiring in the opposite direction—to kill nuclear arms control.

Under President Putin, it became an open secret that Russia was chafing at the INF Treaty's limits. Russia lives in a dangerous neighborhood with intermediate-range missiles all around, particularly in China, which is not bound by INF. Moscow wanted out but did not want to withdraw openly and get the international blame for killing the agreement. Pressure began to mount with US deployment of missile interceptors in eastern Europe under President Obama. Regrettably, Moscow secretly developed, tested, and then deployed "multiple battalions" of up to a hundred land-based cruise missiles (with the catchy name of SSC-8 or 9M729) that are prohibited by INF.[6]

The Obama administration saw this coming in 2014 and wisely chose to stay inside INF and work with the Russians to get them back into compliance. This effort did not succeed by the end of the administration, but at least Obama did not withdraw and thereby give Russia an excuse to openly break from the treaty and build many hundreds of new weapons.

President Trump, however, fell right into Russia's trap. Rather than seek to keep Russia inside INF and thereby at least partially constrained, President Trump handed Putin a free pass to get out of the deal—and have Washington take the blame. Russia is now free to build as many land-based intermediate-range missiles as it wants. This is the worst possible outcome.

Think of it like highway speed limits. The limit is 65, and Russia is speeding at 75. If we throw away the speed limit, Russia can now go 125 with no constraints. How is that better?

Meanwhile, in addition to the new weapons under development to replace the nuclear "triad," the Trump administration is fielding new, "more usable" nuclear warheads for use on strategic submarine-launched ballistic missiles (SLBMs) and is developing a new fleet of sea-launched Tomahawk cruise missiles (SLCMs). And now that the Trump administration is no longer bound by INF, it has started testing a new ground-launched cruise missile and ballistic missile, formerly prohibited by the treaty.[7]

Russia sees these new INF-range missiles as a significant threat, particularly because they have very short flight times from western Europe to Moscow and provide almost no warning of attack. "This is a very serious danger to us," Putin said on February 20, 2019. "In this case, we will be forced, and I want to stress this, we will be forced to envisage tit-for-tat and asymmetric measures." He added that Russia would not only target the launch site of any Europe-based missiles, but—referring to the United States—would target the command center from which they are launched.[8]

Of course, Russia already targets the United States with nuclear weapons, and the United States targets Russia. This is nothing new. What Putin was saying to Americans is that nuclear war will not stay "over there."

## THE ROAD NOT TAKEN

Much like solving a family squabble, instead of starting a new arms race, the Trump administration could have sought to save INF by sitting down with Moscow and putting all the concerns on the table. Russia has its own problems with US actions under INF, such as Washington's deployment of missile interceptor sites in Poland and Romania that could launch offensive cruise missiles.[9] "The United States demonstratively neglected the provisions of the INF Treaty when deploying launchers in Romania and Poland," Putin said. "We don't want confrontation, particularly with such a global power as the US. But Russia will always respond."[10]

Russian ambassador Antonov wrote in 2019, "In 2014, they [the United

States] began to deploy in Europe the Mk-41 vertical launching systems. These launchers are fully suitable, without any substantial modifications, for Tomahawk intermediate-range attack missiles. And this is a clear violation of the [INF] Treaty. Launchers of this kind have already been deployed in Romania and it looks like next year they will be deployed in Poland."[11] In fact, the US test of a Tomahawk cruise missile in August 2019, just after leaving INF, was from a Mk-41 ground-based launcher.

There is an important precedent for walking Russia back from the brink. In the 1980s, the Reagan administration correctly accused Russia of violating the 1972 ABM Treaty. President Reagan worked with Moscow to come back into compliance, and that is what President Trump should have done too.

Some prominent Democrats have gotten behind this approach. "My view is, we have a moral and strategic responsibility to do everything in our power to prevent a new nuclear arms race," Senator Elizabeth Warren said. "And at a minimum, I think that means working with Russia to try to get back to the negotiating table, try to get them back into compliance with the INF Treaty and working on a New START Treaty. This just seems to me to be commonsense arms control and to make America safe."[12]

"The Trump administration is risking an arms race and undermining international security and stability," Speaker Nancy Pelosi (D-CA) said after Trump's announcement. "The administration should exhaust every diplomatic effort and work closely with NATO allies over the next six months to avoid thrusting the United States into a dangerous arms competition."[13]

NATO could play a useful role by declaring that no members will deploy weapons once prohibited by INF as long as Russia does not deploy additional such weapons within range of NATO states. I (Tom) was in the House gallery to see NATO secretary general Jens Stoltenberg tell a joint session of Congress on April 3, 2019, that "NATO has no intention of deploying land-based nuclear missiles in Europe." This is a good start, although initial Trump administration plans call for the deployment of *conventional* missiles in Europe, not nuclear. But such missiles could later be armed with nuclear warheads, so it is not clear that Russia would see a distinction. Both could be a destabilizing threat to Russia's strategic forces.

Some will argue that the United States must force Russia to remove the illegal cruise missiles from eastern Europe. But it is important to realize that Russia's INF violation is primarily a political problem, not a military threat. It does not change the balance of power in Europe and thus, from a military perspective, does not require a response. As Air Force Gen. Paul Selva, then vice chairman of the Joint Chiefs of Staff, told the Senate in July 2017, "We're not restricted from fielding ballistic missiles or cruise missile systems that could be launched from ships or airplanes under the Intermediate Nuclear Forces Treaty; it is specific to land-based missiles." So, any potential military response could be filled by weapons that are not covered by INF, and that already exist, such as US air-delivered bombs that are already in Europe. "Given the location of the specific missile and the deployment, [the Russians] don't gain any advantage in Europe," Selva said.[14]

George Shultz, former secretary of state under President Ronald Reagan, had this to say about Trump's action: "Now is not the time to build larger arsenals of nuclear weapons. Now is the time to rid the world of this threat. Leaving the [INF] treaty would be a huge step backward. We should fix it, not kill it."[15]

Mikhail Gorbachev, who negotiated INF with Reagan, said, "President Trump announced last week the United States' plan to withdraw from the Intermediate-range Nuclear Forces Treaty and his country's intention to build up nuclear arms. I am being asked whether I feel bitter watching the demise of what I worked so hard to achieve. But this is not a personal matter. Much more is at stake. A new arms race has been announced. The I.N.F. Treaty is not the first victim of the militarization of world affairs."[16]

For those keeping score, Putin is to blame for violating the INF Treaty and Trump is to blame for withdrawing from the agreement too quickly, before exhausting options to deal with Moscow's concerns. Together, they have killed INF.

## THE BOLTON FACTOR

Trump's former national security advisor John Bolton, who has long been opposed to arms control, wanted INF dead long before the Russian violations were revealed. In a 2011 article in the *Wall Street Journal*, Bolton wrote that INF was no longer needed, referencing Charles de Gaulle's remark, "Treaties, you see, are like girls and roses: They last while they last."[17]

Bolton argued that INF constrained only Washington and Moscow while other nations, particularly China, remained free to develop intermediate-range missiles. Bolton said the United States should expand the treaty to include other members or withdraw to develop its own prohibited weapons. "The US motto on the INF should be: expand it or expunge it," Bolton wrote.

Three years later, in another article in the *Wall Street Journal*, Bolton no longer suggested expanding INF into a multilateral treaty but simply called for the United States to withdraw in response to the Russian violations that were first reported in 2014.[18]

Bolton, who has rightly earned the reputation of an arms control wrecking ball, also led the Bush administration to withdraw from the Agreed Framework with North Korea in 2002 and the Trump administration out of the Iran nuclear deal in 2018. These moves were major strategic mistakes.

But Bolton's biggest blunder was enacted by President George W. Bush, who in 2002 withdrew from the ABM Treaty with the Soviet Union to deploy limited missile defenses against North Korea that still do not work reliably enough to be useful, and probably never will (see next chapter). Yet the United States has wasted many tens of billions of dollars on the program. Even more important, the demise of the ABM Treaty was the initial and most critical step in the later collapse of other treaties and in restarting the nuclear arms race.

Let us make this point again because it is so important. US defenses against long-range missiles would be completely ineffective against a major Russian attack; they would simply be overwhelmed by numbers. Even an attack by North Korea could overwhelm our defenses if North Korea used a large number of decoys along with their warheads, which seems highly likely. But missile defenses are worse than ineffective; they actually serve to *increase* the threat we face from Russia and China, as those countries oppose additional arms reductions (Moscow) or seek to increase their forces (Beijing).

Bolton is gone from the Trump administration, but the treaty wrecking may not be over. The New START Treaty, negotiated by the Obama administration and signed by the United States and Russia in 2010, is the last major agreement still in force limiting nuclear arms, and it expires in February 2021. It can be renewed for five years but only if Washington and Moscow agree. Russia claims that it wants to extend the treaty. Trump may do nothing or actively oppose extension.


## SAVING NEW START


In a January 2017 phone call with Russian president Putin, President Trump reportedly did not know what New START was but denounced it as a bad Obama administration deal.[19] Asked about the report, then White House spokesperson Sean Spicer first avoided the question but later denied that Mr. Trump was uninformed about the treaty.[20] Then national security advisor John Bolton confirmed our suspicions when he said in July 2019 that New START is "unlikely to be extended. Why extend a flawed system just to say you have a treaty? We need to focus on something better. And we will."[21]

Unfortunately, the fallout from the Bush administration's ABM Treaty withdrawal almost two decades ago is still being felt today and is impacting New START. After the US withdrawal in 2002, President Putin quietly decided to take precautionary measures against a major expansion of US missile defenses by developing new offensive weapons. Sixteen years later, in March 2018, Putin gave a major speech and reported on his progress. He described a new nuclear-armed, unmanned, underwater drone known as Status-6; a new long-range, nuclear-powered cruise missile called Skyfall; new hypersonic weapons; and more. A key feature shared by all of these new weapons is their ability to evade US missile interceptors. Russia deployed a new hypersonic nuclear warhead, called "Avangard," on existing ICBMs in December 2019. The United States is expected to follow suit by 2022.[22]

"In 2002, the United States has decided to withdraw from ABM Treaty, and in that year, we made it clear to everybody that we will face bad consequences of such a decision," said Russian ambassador Antonov in March 2019. "And, on first of March last year, President Putin announced about new Russian arms and armaments that we developed. Some politicians and military officers were surprised to see what kind of weapons we created, but we tried to explain that it was just only simple answer from Russian side to the decision by the United States to withdraw from ABM Treaty."[23]

One of Russia's new weapons, the nuclear-powered cruise missile, is a particularly terrible idea, because it is basically a flying nuclear reactor. If it crashes during a test, you have another Chernobyl. Russia conducted a test of the Skyfall cruise missile in 2018, sending it crashing into the White Sea, where it remained until August 8, 2019. When Moscow recovered the missile, the nuclear reactor exploded at a test site near Nyonoksa, killing five employees of Rosatom, the Russian atomic energy corporation, and causing a brief rise of the radiation level, according to local authorities.[24]

Russia is fully responsible for this reckless activity, but we must also point out that Moscow was developing this missile in response to the US ABM Treaty withdrawal. This is the kind of unintended consequence that such misguided actions can cause.

In addition to being dangerous, some of the new Russian weapons are not explicitly covered by New START. Now, some in the United States do not want to renew New START without accounting for these new weapons in some way, which would entail a new treaty. And if the United States wants to reopen the treaty instead of extending it "as is," Moscow will have some demands as well. For example, Antonov has called for addressing missile defense, nonnuclear strategic weapons, weapons in space, and cybersecurity, among other issues.

The compromise solution should be a "clean" extension of New START for five years, during which time the two sides can work on a new follow-on agreement that could take account of these new developments. But it is essential that these new issues do not prevent extension of the treaty in 2021.

Russian ambassador Antonov wrote in April 2019, "We hope that the New START will not suffer the same fate as the INF Treaty. It expires in 2021. On many occasions we have declared our readiness to discuss the possibility of its extension for another five years. Washington, however, still cannot give a definite answer."[25]

Instead of simply extending New START, President Trump has directed his administration to seek a new arms control agreement with Russia and

China that should include "all the weapons, all the warheads, and all the missiles."[26]

China has rejected engaging in these proposed talks, on the grounds that it has an arsenal that is significantly smaller than the others. Administration officials appeared before the Senate Foreign Relations Committee in May 2019 to inform Congress on its New START policy, but their testimony only raised more questions.

Ranking Democrat Robert Menendez (D-NJ) asked then under secretary of state Andrea Thompson, "If Russia is in compliance, do you believe it is in the nation's best interest to extend New START?"

She replied, "We're engaged in an interagency process" and "it is too soon to tell."

Menendez: "If New START expires, could Russia target the United States with hundreds or perhaps thousands of additional nuclear warheads?"

Thompson: "That is a great question for Russia, Senator."

Menendez: "No, that is a great question for you . . . I am not asking Russia about our national defense, I am asking you!"

Thompson: "That's a hypothetical, Senator, and I am not going to answer that."[27]

In fact, there is nothing hypothetical about this. If New START expires, Russia could add hundreds of additional nuclear warheads onto its long-range missiles, increasing the nuclear threat to the United States. Of course, the United States could also add warheads to its missiles.

Senator Menendez noted, "This new initiative must not serve as an excuse for suddenly withdrawing from another international agreement. If new agreements can be reached, they should add, not subtract from, our existing arms control architecture."[28]

Abandoning New START would be a tragic error that would throw gasoline on the arms race fire. New START has served the United States well, and there are no indications of Russian violations. It deserves to be renewed. It limits the number of nuclear weapons that Russia can aim at the United States, and it gives Washington confidence that Moscow will not expand this arsenal quickly. Without New START, those assurances disappear. New START is also the vehicle for the last remaining dialogue we have with Russia on nuclear weapons. Indeed, one of the most important reasons for strategic arms treaties is to maintain a continuing dialogue on these issues that affect the survivability of both of our countries.

In an interview with former secretary of defense James Mattis, we asked him if he supported the extension of New START. He said, "Yes, no doubt. Arms control has got to be a part of our deterrence strategy. This highlights why the damage caused by Russia's violation of the INF Treaty is so serious, casting doubt as it does over the entire arms control effort."[29]

Former STRATCOM commander John Hyten also supports New START. He testified that the treaty "gives me two very important things. Number one, it puts a limit on the basics of their strategic force. So, I understand what their limits are and I can position my force accordingly so I can always be ready to respond. And maybe as important, it also gives me insight through the verification process of exactly what they're doing and what those pieces are. Having that insight through my forces and our partners is unbelievably important for me to understand what Russia is doing."[30]

If New START is killed, it would be the first time there have been no speed limits on US–Russian nuclear arsenals since 1972. We would be back in the Wild West. Without New START, we would be careening, with nuclear weapons, along a dangerous mountain road, with no information as to what lies around the next bend.

## FORCE REDUCTION, THEN AND NOW

Over the last fifty years, the United States and Russia have dramatically reduced the number of nuclear weapons in their arsenals. The US nuclear stockpile peaked at 31,255 warheads in 1967 and has come down ever since.[31] In certain periods of time, the force reductions were large. As the Cold War was ending during the George H. W. Bush administration, from 1989 to 1993, the US nuclear stockpile dropped by 50 percent, the most rapid nuclear arsenal reduction in US history. During the George W. Bush administration, from 2001 to 2009, the stockpile was cut in half again.

Yet while the numbers have come down, we still have enough to destroy the world. Scientists estimate that, in addition to tens of millions of immediate deaths, a US–Russian war using current nuclear arsenals could cause below-freezing temperatures over much of the northern hemisphere during summer, possibly lasting years.[32] Despite the Trump administration's attack on arms control, it is imperative that the arms reduction process continue.

US military leaders have determined time and again that nuclear stockpiles are larger than needed to maintain the security of the United States and its allies and friends. These arsenal reductions have encouraged corresponding reductions by Russia, thereby lowering the nuclear threat from the only nation capable of "turning the United States into radioactive ash."[33] Moreover, US and Russian reductions have helped build international support for stopping the spread of nuclear weapons to other states or terrorist organizations, a significant threat to US security.

Enhancing US national security by verifiably reducing superpower nuclear arsenals—a counterintuitive idea to some—has a long bipartisan tradition. US presidents beginning with Lyndon B. Johnson have pursued and signed bilateral agreements mandating verifiable limits and reductions in US and Russian nuclear stockpiles. Presidents Richard Nixon, Gerald Ford, Jimmy Carter, Ronald Reagan, George H. W. Bush, Bill Clinton, George W. Bush, and Barack Obama all contributed to reducing the nuclear threat through the negotiation of nuclear arms control agreements with the Soviet Union and later with Russia.

President Richard Nixon and General Secretary Leonid Brezhnev took the first step to cap US and Soviet nuclear ballistic missile forces and defenses with the Strategic Arms Limitation Talks Agreement (SALT I) and the Anti-Ballistic Missile (ABM) Treaty in 1972.

The follow-on SALT II Treaty, signed by President Jimmy Carter and Brezhnev in June 1979, was submitted to the US Senate for ratification shortly thereafter. But Carter removed the treaty from Senate consideration in January 1980, after the Soviet Union's invasion of Afghanistan. Nevertheless, the United States and the Soviet Union voluntarily observed the SALT II limits. By this time, the US arsenal had been reduced to about 24,000 warheads.

President Ronald Reagan began talks toward the Intermediate-Range Nuclear Forces (INF) Treaty, which he and Soviet leader Mikhail Gorbachev signed in 1987. As Reagan said in 1986, "It is my fervent goal and hope . . . that we will someday no longer have to rely on nuclear weapons to deter aggression and assure world peace. To that end, the United States is now engaged in a serious and sustained effort to negotiate major reductions in levels of offensive nuclear weapons with the ultimate goal of eliminating these weapons from the face of the earth."[34]

Under INF, the two nations agreed to eliminate their stocks of medium-range, nuclear-capable, land-based missiles. It was the first arms control treaty to abolish an entire category of weapons systems and established unprecedented procedures to verify firsthand that missiles were actually destroyed. The US Senate gave its advice and consent to the INF Treaty in 1988.

Meanwhile, Reagan and his team pursued negotiations on the START Treaty with the Soviets. Under START, President Reagan proposed major reductions, not just limitations, in each superpower's stockpile of long-range missiles and bombers. The START I Treaty was signed by President George H. W. Bush and Gorbachev in 1991, and the US Senate gave its advice and consent in 1992.

In late 1991, the Soviet Union broke up and created the independent states of Russia, Belarus, Kazakhstan, and Ukraine. The most significant danger emanating from the former Soviet Union was the loss of control of its nuclear stockpile.

President George H. W. Bush responded with his bold Presidential Nuclear Initiatives (PNIs) in September 1991, which led to the removal of thousands of US tactical nuclear weapons from forward deployment. Days later, Moscow reciprocated, reducing the risk that these weapons would fall into the wrong hands. No formal treaty was ever negotiated or signed, nor did the Bush administration seek the approval of Congress. Under the PNIs and subsequent actions, the United States unilaterally reduced its stockpile of nonstrategic warheads by 90 percent.[35]

But the breakup of the Soviet Union also created a problem with *strategic* (long-range) nuclear weapons. When the Soviet Union dissolved, there were more than 4,000 nuclear warheads on strategic weapons based in Ukraine, Kazakhstan, and Belarus, and these deadly devices became the property of these three nations even though they had no provisions for dealing with them. Senators Sam Nunn (D-GA) and Richard Lugar (R-IN) saw the danger of this "loose nukes" problem and pushed through, with bipartisan support, the Nunn-Lugar program to deal with it. When I (Bill) became secretary of defense, executing the Nunn-Lugar program was my highest priority. I redirected the necessary funding from other programs in the Defense budget and put together a crack team headed by Ash Carter (who would later become defense secretary under Obama), whose goal was to use the authority and funding from this program to dismantle all nuclear weapons in those three states by the end of President Clinton's first term in office.

We achieved that goal, but only through a high level of cooperation from Russia, Ukraine, Belarus, and Kazakhstan. The good will and experience that had already been built up through the robust arms control negotiations just prior to that period were essential to the success of the Nunn-Lugar program. An interesting footnote to the Nunn-Lugar program was that all of the highly enriched uranium that was taken from those dismantled weapons was blended down and used to power commercial nuclear reactors in the United States, providing power for our electric grid. The Nunn-Lugar program averted a nuclear disaster whose dimensions we can only imagine, but its success absolutely depended on prior arms control treaties.

The START II Treaty was signed in early 1993 but was not taken up by either the Senate or the Russian Duma. In early 1996, I (Bill) urged the Senate to bring it up and ratify it, which they did. Later that year, I testified to the Duma on why they should also ratify START II, but I left office before they voted on it. In 2000, the Duma linked the fate of START II to the continuation of the 1972 Anti-Ballistic Missile (ABM) Treaty. Following the George W. Bush administration's withdrawal from the ABM Treaty in June 2002, the Duma rejected START II.

In March 1997, Clinton and Yeltsin agreed to begin negotiating START III, which would have reduced each side to 2,000 to 2,500 deployed strategic warheads by December 31, 2007. Unfortunately, discussions bogged down over distinctions between strategic and theater-range missile intercep-tors under the ABM Treaty, and START III was never concluded, another casualty of US withdrawal from the ABM Treaty.

In May 2002, George W. Bush and Vladimir Putin signed the Strategic Offensive Reductions Treaty (SORT or Moscow Treaty), which limited both sides' strategic warheads to 1,700 to 2,200. The US Senate gave its advice and consent to SORT in 2003.

It is worth noting that President Bush initially set out to reduce US forces without a formal agreement. As he said in 2001: "We don't need an arms control agreement to convince us to reduce our nuclear weapons down substantially, and I'm going to do it."[36]

President Bush ultimately agreed to submit SORT to the Senate in part because Russia wanted a treaty, even if it was a very simple one (a single page of text) with no verification measures. Had Russia not wanted a formal agree-ment, Bush would likely have reduced US nuclear weapons without a formal agreement, as his father did before him.

The SORT Treaty relied indirectly on the verification mechanisms of START I, so the United States and Russia both wanted to negotiate a new bilateral agreement before START I expired in 2009. They did not quite make it but did achieve a new treaty one year later.

In April 2010, Presidents Obama and Dmitry Medvedev signed the New START Treaty to limit each side to 1,550 deployed, treaty-accountable stra-tegic warheads. The Senate gave its advice and consent to the agreement in December 2010, and it entered into force on February 5, 2011. Both parties met the central limits of the treaty by February 2018. The treaty will expire in February 2021.

When New START was signed, little significance was attached to the renewal date, because the pact was seen as an interim step to a follow-on treaty with significantly deeper cuts. As President Obama said in Berlin in June 2013, "We can ensure the security of America and our allies, and maintain a strong and credible strategic deterrent, while reducing our deployed strategic nuclear weapons by up to one-third," from 1,550 New START–accountable deployed strategic warheads to about 1,000.[37]

## ARMS REDUCTIONS STILL MAKE SENSE

Today's most pressing nuclear security threats to the United States are accidental war, nuclear terrorism, and proliferation. Excessive US nuclear forces have no meaningful role to play in this regard. The United States needs to sustain a strong international coalition to secure nuclear materials across the globe and turn back nuclear programs in Iran and North Korea, and continued US and Russia arms reductions are essential to these goals.

In addition, by clarifying their intentions to achieve further nuclear arms reductions and taking steps in that direction, US leaders can put more pressure on China to exercise greater restraint and engage more actively in nuclear risk-reduction initiatives.

It remains in the US interest to reduce Russia's arsenal of nuclear weapons, despite the welcome fact that the threat of intentional nuclear attack from Moscow has decreased. Russian arsenal reductions can still decrease the consequences of possible accidental missile launches and help serve the goal of providing better security for and ultimately eliminating weapons-usable materials.

Former Senator Carl Levin (D-MI), then chair of the Senate Armed Services Committee, said in June 2012: "I can't see any reason for having as large an inventory as we are allowed to have under New START, in terms of real threat, potential threat." He added, "The more weapons that exist out there, the less secure we are, rather than the more secure we are."[38]

A key reason to continue the US–Russian arms control process is to strengthen the international coalition against proliferation. This is where some of the greatest future threats to US security lie. Excessively large arsenals do not stop proliferation, yet arsenal reductions can translate into greater global support for US nonproliferation efforts.

Finally, fiscal pressure on the defense budget makes it unwise to maintain any military program that is larger than it needs to be. A dollar wasted on excess nuclear weapons is a dollar lost to preventing terrorism or proliferation. In 2003, then secretary of state Colin Powell noted: "We have every incentive to reduce the number [of nuclear weapons]. These are expensive. They take away from soldier pay. They take away from [operation and maintenance] investments. They take away from lots of things. There is no incentive to keep more than you believe you need for the security of the Nation."[39]

Two arguments that are often made against lowering the US nuclear arsenal are that it would encourage China to build up and would cause such worry to our allies that they may decide to build their own nuclear forces. Neither argument holds water.

For decades now, China has been content with a much smaller nuclear arsenal than that of the United States or Russia. Beijing has a total estimated stockpile of fewer than three hundred warheads. Even after reducing to a thousand deployed strategic warheads, the United States would still enjoy a three-to-one advantage. China poses no roadblock to continued nuclear reductions at this time. And if the United States and Russia reduce their nuclear forces, Washington and Moscow will be in a better position to reach an understanding with China about limiting the further growth of its arsenal.

On the other hand, maintaining unnecessarily large US and Russian nuclear forces, combined with increasingly large (though unreliable) US long-range ballistic missile defenses, could push China to accelerate its efforts to increase the size and capabilities of its strategic nuclear force.

Some critics claim that further US nuclear force reductions would drive allies that depend on the so-called US nuclear "umbrella" to reconsider their nonnuclear weapon status and seek their own arsenals.

Such concerns are unfounded given the unmatched retaliatory potential of even a few hundred US nuclear weapons, as well as the overwhelming superiority of US conventional forces. Moreover, for a nonnuclear state, such as South Korea or Japan, to openly build a nuclear arsenal would be a dramatic renunciation of its commitment not to do so under the Nuclear Non-Proliferation Treaty. The political costs of such a decision would be huge.

Furthermore, rather than express opposition to more nuclear force reductions, US allies in Europe and Japan have consistently and repeatedly called on the United States and Russia to "continue discussions and follow-on measures to the New START to achieve even deeper reductions in their nuclear arsenals towards achieving the goal of a world free of nuclear weapons" and they "urge those not yet engaged in nuclear disarmament efforts to reduce their arsenals with the objective of their total elimination."[40]

Today, it is clear that the United States can maintain a credible deterrent at significantly lower levels of nuclear weapons than we currently have. There is no reasonable justification today for such high numbers. Further reductions to the US nuclear stockpile would bring a variety of benefits, including the prospect of a smaller Russia arsenal, a stronger international coalition against nuclear terrorism and proliferation, and billions of dollars that could be saved or spent on higher priority defense needs. Nuclear arsenal reductions have made sense to seven presidents over five decades. They still make sense today.

But getting there will involve overcoming some difficult challenges, such as limiting national missile defenses—the third rail of nuclear security politics. As long as Russia sees US missile defense as a threat to its ability to retaliate after a US first strike, it will refuse to negotiate major new arms reductions without limits on US defenses. And as long as the US Congress remains under the mistaken belief that these systems can be made to work effectively, it will refuse to limit national defenses. The missile defense dilemma is discussed in detail in the next chapter.

President Kennedy and Maj. Gen. Clifton accompanied closely by a military aide carrying the nuclear football.

President Nixon speaks with military leadership on the tarmac while the nuclear football follows close behind.

President Ford in conversation with the military aide carrying the nuclear football as he departs from the White House.

A military aide bearing the nuclear football walks alongside a smiling President Reagan.

President George H.W. Bush finishes a jog with his security detail and the nuclear football in tow.

President Clinton, flanked by a military aide with the nuclear football, greets military officials while boarding Air Force One.

President George W. Bush and First Lady Laura Bush wave to onlookers as they board Air Force One. A military aide carrying the nuclear football waits out of sight nearby.

President Trump departs from the CIA after delivering a speech, followed closely by a military aide with the nuclear football.

# CHAPTER 8

# THE MISSILE DEFENSE DELUSION

*But let no one think that the expenditure of vast sums for weapons and systems of defense can guarantee absolute safety for the cities and the citizens of any nation. The awful arithmetic of the atomic bomb does not permit such an easy solution.*
—PRESIDENT EISENHOWER[1]

O n March 25, 2019, the US Missile Defense Agency (MDA) conducted a test high over the Pacific Ocean. MDA launched an unarmed target missile toward the United States from a test site on Kwajalein Atoll in the Marshall Islands. Then, over 4,000 miles away, the Pentagon launched a "salvo" of two interceptor missiles from Vandenberg Air

Force Base, just north of Los Angeles, California, to shoot the target missile down.

Once in outer space, the target missile from Kwajalein released a mock warhead traveling at 22,000 miles per hour. Heading the other way, the two interceptor missiles from Vandenberg released "kill vehicles" to hit the mock warhead. These interceptors do not carry explosives of any kind and must destroy the target by force of impact—they must hit it dead-on, known as "hit to kill."

The test was a "success" in the sense that the first interceptor missile, called a ground-based interceptor (GBI), hit and destroyed the mock warhead. The second GBI looked to see if the target was destroyed and, seeing it was, hit a piece of debris.

"The system worked exactly as it was designed to do, and the results of this test provide evidence of the practicable use of the salvo doctrine within missile defense," said MDA director Lt. Gen. Samuel A. Greaves after the test. "The Ground-based Midcourse Defense system is vitally important to the defense of our homeland, and this test demonstrates that we have a capable, credible deterrent against a very real threat."[2]

As then-Under Secretary of Defense for Policy John Rood put it after the test, "The United States is protected by the Ground-based Missile Defense system."[3]

The average person reading such statements would think, "Great. Missile defense works. I don't have to worry about a missile attack from North Korea or Russia. If they launch an attack, we can stop it."

We wish this were the case. We would like nothing more than to share Greaves or Rood's confidence that the US defense against long-range missiles really works. But, sadly, it is just not true. Hitting a "bullet with a bullet" may be an impressive technical achievement, but this does not demonstrate that "the United States is protected by the Ground-based Missile Defense system." Not even close.

These are not the first overstatements of what the Ground-based Midcourse Defense (GMD) system can do. In fact, you would be hard-pressed to find a senior Trump or Obama administration official, or a member of Congress from either party, that would tell you the honest truth: *we do not know whether the GMD system would work against realistic threats because the United States has never tested this system against such threats.*

Here is what we believe: based on the testing record to date, no rational US president could be confident that the current US system would stop a realistic attack from North Korea, much less China or Russia. The system has only "worked" about half the time in tests, which have been intentionally dumbed down to allow the system to succeed. Most important, the tests have not included the kinds of decoys that we must assume will be used by Russia, China, and, yes, North Korea. And because the GMD system conducts its intercepts in outer space, it is especially vulnerable to decoys. (See the section later in this chapter titled "Why Anti-Ballistic Missiles Don't Work.")

This is why the US Government Accountability Office has reported that the tests have been "insufficient to demonstrate that an *operationally useful* [italics ours] defense capability exists."[4]

What is going on here? The history of the US missile defense program is one of politics and defense spending trumping science. Politicians and defense officials have misinformed Americans that missile defense works, and it is complicated (and politically awkward) to explain why it does not. Defense contractors and their employees are strong supporters. Billions of dollars are at stake. Meanwhile, politicians who may question inadequate defenses open themselves to attacks that they "do not want to protect America."

Some might say, "No harm done. The government has wasted more money on worse things." We do not agree. The danger here is quite real. First, the more money we as a nation spend on ineffective missile interceptors, the more we convince ourselves that they will work. How else can we justify spending so much money? But this overconfidence is dangerous. If President Trump believes he can intercept a missile attack, he may be more likely to escalate a conflict. This is how nations stumble into unintended wars. In October 2017, in reference to the GMD system, President Trump told Fox News' Sean Hannity, "We have missiles that can knock out a missile in the air 97 percent of the time, and if you send two of them it's gonna get knocked out." We have no idea if Trump really believes this, but if he does, it is highly dangerous.

If Trump believes his anti-missile interceptors would provide a workable defense, then he might push a crisis beyond the danger point on the mistaken belief that any resulting missile attack would not get through to American cities. Or Trump might even launch a preemptive strike against North Korea, on the same faulty logic. "This misperception could be enough to lead the United States into a costly war with devastating consequences," wrote Ankit Panda, an adjunct senior fellow at the Federation of American Scientists, and Vipin Narang, an associate professor of political science at MIT.[5]

The second danger of an ineffective anti-missile system is that, even if Russian leaders do not really believe it will work, they cannot afford to assume it won't. So, Russia will hedge and plan for the worst case: that the US can mount a defense against ballistic missiles. This has already led Russia to develop several new systems to attack the United States, including missiles that have a long enough range to attack from the south, thereby evading US missile defense systems. Putin proudly described these new systems to the Russian people in his State of the Federation speech in March 2018.

A recently declassified CIA bulletin from around 2000 found that:

> Moscow continues to perceive US plans for even a limited missile defense system as undermining its strategic retaliatory capability. . . . Moscow is concerned that its declining strategic nuclear forces could no longer survive a first strike with enough missiles left to overcome US missile defenses, undermining its ability to deter a US attack.[6]

This CIA statement says a lot. Amazingly, Russia really fears a US first strike. Moreover, Russia is apparently worried that it cannot deter a first strike that might leave it with so few nuclear forces for retaliation that those missiles could be stopped by even a limited US missile interceptor system.

This unfounded fear is based on two false assumptions. First, it is highly unlikely that Washington would launch an unprovoked first strike. That is not in the cards for the simple reason that, as we have argued, no rational US president would take that risk with the fate of the nation at stake.

Second, it is also false that US missile defenses would be effective against Russian missiles, even in the case of a smaller retaliation. We believe Russian scientists know this to be false but apparently have been unable to convince their political leaders.

When you put these two misguided beliefs together, you get a Russian government that is suspicious of US intentions and resistant to nuclear arms reductions. That resistance was managed in the case of the New START Treaty by limiting its duration to ten to fifteen years. The preamble acknowledges the "interrelationship between strategic offensive arms and strategic defensive arms" and that "current strategic defensive arms do not undermine the viability and effectiveness of the strategic offensive arms of the Parties." Even so, Russia made a unilateral statement that it could potentially withdraw from New START if the United States were to deploy missile interceptors in large numbers. When signing New START, Russian president Medvedev said the treaty "can operate and be viable only if the United States of America refrains from developing its missile defense capabilities quantitatively or qualitatively."[7]

And so, when President Obama in 2013 offered to negotiate another treaty with Russia to reduce nuclear forces further than New START, Moscow said *nyet*. A major factor in Moscow's refusal was its concern about expanding US missile defenses. As things stand now, Washington is letting its support of missile defenses—that do not protect the United States—prevent the reduction of Russian nuclear weapons through arms control agreements—which do.

## HOW WE GOT HERE

Back in the 1960s, US arms control experts knew quite well that if either side built sizable missile defenses, it would be harder to reduce offensive forces. The simple reason for this is that if your adversary is building defenses, the best answer is to build more and better offenses. Think of castles. As castle walls got higher, ladders got taller. As walls got thicker, cannon balls got heavier. As moats got wider, catapults added range. The answer to better defenses has always been, and still is, better offenses. The sword is always more powerful than the shield, even in the atomic age.

So, before US and Russian negotiators could limit offensive nuclear weapons, they had to limit the defenses against them. They did this with the 1972 ABM Treaty, the preamble of which states that "effective measures to limit anti-ballistic missile systems would be a substantial factor in curbing the race in strategic offensive arms and would lead to a decrease in the risk of outbreak of war involving nuclear weapons."[8]

The ABM Treaty prohibited the deployment of nationwide missile defenses by limiting each side to a hundred interceptor missiles at one location (reduced from two) in each nation to protect their capitals or ICBM fields. The Soviet Union chose to put its interceptors around Moscow, which are still in place today (and are regarded to be ineffective and easily overwhelmed). The United States opted to defend an ICBM site in North Dakota with a system called Safeguard but shut it down after a few months in October 1975 because Congress concluded it was too expensive and unreliable. The treaty also banned sea-, air-, and space-based anti-missile systems.

The fact that the treaty allowed even a hundred interceptors was a seed of the pact's demise. It meant that powerful factions in both countries believed in—or stood to profit from—missile interceptors. And the treaty allowed continued research into better defenses. As President Reagan would show us, *defending* the nation is a powerful idea, more powerful than deterrence or mutually assured destruction.

The uneasy compromise of the ABM Treaty lasted until March 1983, when President Reagan began calling for his space-based Strategic Defense Initiative (SDI), lampooned as "Star Wars," that would render Russian nuclear weapons "impotent and obsolete." This plan would have deployed a nationwide defense and weapons in space, violating the ABM Treaty. Reagan argued that research and testing could take place within the treaty, but Gorbachev regarded any testing outside the laboratory as a violation. Tragically, this dispute blocked agreement on nuclear disarmament at the 1986 Reykjavik Summit in Iceland.

Ten years later, after spending tens of billions of dollars on X-ray lasers, directed-energy weapons, particle-beam weapons, space-based kinetic interceptors, and "Brilliant Pebbles," the Pentagon was forced to conclude that none of these concepts would work. The idea of a massive defense against hundreds of incoming warheads was dead.

By the time President Bill Clinton was considering nationwide missile defense deployment in 2000, the threat du jour was not Russia but North

Korea, and the proposed defense was made up of ground-based intercep-tors to knock out one or two warheads, not hundreds. The genesis of this threat was the August 1998 test of Pyongyang's Taepodong 1 three-stage space-launch vehicle. This missile flew over Japan and woke the world up to North Korea's dangerous potential. The test demonstrated that the North had gained expertise in the ability to launch missiles with multiple stages, a key step in the development of ICBMs.

A national intelligence report in 1999 estimated that North Korea could have tested a more capable missile that year but did not for political reasons. According to the intelligence, a three-stage Taepodong 2 could deliver a small nuclear warhead anywhere in the United States. I (Bill) was asked by Pres-ident Clinton to conduct a policy review on North Korea and seek to stop the North's nuclear and ICBM programs. I did conduct such a review, jointly with colleagues from South Korea and Japan, and in 1999 I took my team to Pyongyang to negotiate with North Korean officials. We offered them vari-ous incentives (including US recognition and an official end to the Korean War) for a verifiable ending of their nuclear and ICBM programs. They were very positive and sent their senior military leader to Washington for final negotiations. That led to an agreement that was to be signed by President Clinton and Kim Jong Il. But before the signing could take place, the incom-ing Bush administration cut off all dialogue with North Korea, apparently believing that strong pressure on them would bring about regime collapse. The hoped-for collapse never occurred; what did occur was the creation of a medium-size North Korean nuclear arsenal with ballistic missiles to carry them to their targets.

While it was almost laughable to argue that US missile interceptor tech-nology could provide an effective defense against many hundreds of advanced Russian ICBM warheads, it was a different story when it came to North Korea. Missile defense proponents argued that they could handle Pyong-yang because the missile threat presumably would be much smaller and less sophisticated.

The emerging North Korean missile threat created considerable political pressure on President Clinton to take decisive action. In a key test of missile defense technology in October 1999, the interceptor hit a mock warhead, and the Defense Department used this as proof that the idea was workable. The timing was crucial, as Clinton was expected to make a decision by the fall of 2000 on deploying the system.

A few weeks after the test, while working at the Union of Concerned Scientists, I (Tom) was given closely held information that the test was not all it was claimed to be. It turned out the interceptor had hit its target only after a series of technical errors had caused it to drift off course. The kill vehicle had initially homed in on a decoy balloon rather than the mock warhead, indicating that the system could easily be fooled.

I gave this information to the *New York Times*, which ran a front-page story on January 14, 2000: "Antimissile Test Viewed as Flawed by Its Opponents."[9] The article said that "the decoy was meant to represent the sort of measures that hostile countries might employ to confuse a national missile defense. But without the large, bright balloon, which in this case happened to be drifting near the smaller and dimmer warhead in the interceptor's field of view, the test last October might not have succeeded, these critics say."

"What this says to me is, if that balloon hadn't been there, then they wouldn't have hit the target," I told the *Times*. "They got lucky."

The article reinforced the notion that the Pentagon was not being straight with the public about the test results. "Congressional investigators found, years after the fact, that tests of hit-to-kill systems in 1984 and 1991 had been partly rigged to make the systems appear more effective than they actually were," the *Times* reported.

In this case, the *Times* found that the kill vehicle got lost but "saw the Mylar balloon in the corner of its field of view. Then, using the balloon as a reference point, the vehicle continued searching for the smaller and cooler warhead and eventually found it nearby, in time to home in on it and destroy it."

Here is the key question: How did the interceptor know in advance that the real target was "smaller and cooler"? The Pentagon had told it so, which made the test unrealistic. The interceptor had been given the right answers before the test. In an actual attack, the potential number of decoys scattered throughout space could be quite large, and the warheads could be made to look like the decoys. There would be no way of knowing in advance whether the real targets would have significantly different sizes or temperatures than the decoys.

In September 2000, partly based on these testing problems, President Clinton made the wise decision not to deploy a limited system to intercept a small number of potential North Korean missiles. He said his decision was based on four criteria: the readiness of the technology, the impact on arms control, the cost of the system, and the threat. Clinton decided that he did not yet have "enough confidence in the technology and the operational effectiveness . . . to move forward to deployment." He noted that there were unresolved questions about countermeasures, or efforts like decoys to "confuse the missile defense into thinking it is hitting a target when it is not."[10]

## BUSH SCRAPS ABM TREATY

On September 11, 2001, everything changed. The terrorist attacks on the World Trade Center and the Pentagon completely transformed the political context for missile defense. It became much more difficult to resist military programs, budgets, and operations that would have been more rigorously debated before. In late 2001, citing a potentially growing threat from rogue nations and terrorism, the George W. Bush administration announced that the United States would withdraw from the ABM Treaty. Soon after, the administration announced plans to rush ahead with an untested missile defense system. Bilateral limits on missile defenses were history.

"Throughout my administration, I have made clear the United States will take every measure necessary to protect our citizens against what is perhaps the greatest danger of all—the catastrophic harm that may result from hostile states or terrorist groups armed with weapons of mass destruction and the means to deliver them," Bush said in 2002.[11]

The Bush administration began fielding the Ground-based Midcourse Defense system in 2004, and today the system is composed of forty-four interceptor missiles in Alaska and California, intended (but unable) to counter a possible long-range missile attack from North Korea. The Department of Defense's current plan is to increase the number of interceptors to sixty-four by 2023 and possibly add more soon thereafter to reach a total of a hundred. The program has so far cost more than $40 billion.[12]

Defense Secretary Donald Rumsfeld initially called the system "limited" but "better than nothing." "I think that anyone who thinks about it understands that if you're at the leading edge of technology, you're going to learn and gain knowledge both by your successes and also by your failures," Rumsfeld said.[13]

Rumsfeld's remarks may make sense for a prototype system that is under development, but not for a system deployed in the field. In its rush to deploy, the Bush administration locked in a flawed technology, and we are still paying the costs. Senator Dick Durbin (D-IL) said in June 2014 that the "design, engineering, and reliability problems . . . were largely caused by the rush to field this system without properly testing [it]. We are now paying dearly for some of those decisions."[14]

Once President Bush deployed a system in the field, the missile defense debate was forever altered. The burden of proof shifted. Before Bush, the burden was on those calling for deployment to make the case that the system would be effective, affordable, and not undermine arms control. Now the burden was on those opposed to deployment to argue why the system should be removed. And the bureaucracy and budgets committed to missile defense had been unleashed. To those who were advocating for adequate testing of missile defenses before fielding, Bush's deployment of an ineffective system was devastating. The untested fielding and the related scrapping of the ABM Treaty were the most damaging things any president has done to the prospects for future US–Russian arms control.

The damage could have been short-lived if President Obama, who inherited the system from Bush, had pulled it down and sought to recreate the ABM Treaty. But it was too late.

At first, Obama tried to keep missile defense at arm's length. As a candidate in 2008, Obama criticized Bush's "haste to deploy missile defenses."[15] Once in office, the Obama administration tried to shift away from national defenses to regional ones, particularly in Europe. But Obama continued to fund Bush's GMD system because canceling it would have led to a huge political fight with Republicans and doomed chances to ratify New START. And once your own administration is paying billions of dollars for something, it is hard to argue that you don't think it will work. And just like that, the deeply flawed GMD system had won bipartisan support, and few dared to question it.

Soon the Obama administration was trapped in its own missile defense rhetoric. In the face of North Korean nuclear and missile tests, the administration expanded the Bush system by roughly 50 percent at the cost of about $1 billion. This made no more sense than Bush's decision to prematurely field the system in 2004. "The idea of deploying 14 more of the existing . . . interceptors at Fort Greely in Alaska, as proposed by the Obama administration last year, would be throwing good money after bad," said Philip Coyle, former director of weapons testing for the Defense Department. "We need to make sure we have a system that works, not expand a system we know to be deeply flawed."[16]

As for the threat it is intended to counter, Adm. James Winnefeld, then vice chairman of the Joint Chiefs of Staff, said in May 2014 that North Korea knows it "would face an overwhelming US response to any attack." As Winnefeld implied, it is not *missile defense* that would stop a missile attack from North Korea; it is *deterrence*.[17]

## MISSILE DEFENSE GOES TO EUROPE

Soon after taking office, the Obama administration sought to remove a major irritant in relations with Russia: the Bush administration's plan to field long-range missile interceptors in eastern Europe to defend the United States against ICBMs from Iran. As it was, there were no ICBMs in Iran, and Moscow saw the US plans as a clear threat to Russian ICBMs. This became a roadblock to Russian support for New START.

So, on September 17, 2009, Obama announced that he would cancel Bush's plan for ten interceptors in Poland and a radar in the Czech Republic and replace it with a different system. Instead of deploying long-range interceptors that were sure to upset Moscow, Obama would deploy shorter-range interceptors at first that could become more capable if the Iranian missile threat grew. Mr. Obama said that the new system "will provide stronger, smarter and swifter defenses of American forces and America's allies" to meet a changing threat from Iran.[18]

The decision drew quick Republican rebuke. "Scrapping the US missile defense system in Poland and the Czech Republic does little more than empower Russia and Iran at the expense of our allies in Europe," said Representative John A. Boehner of Ohio, the House Republican leader. "It shows a willful determination to continue ignoring the threat posed by some of the most dangerous regimes in the world."[19]

But the shift drew guarded praise in Moscow. President Medvedev said, "We appreciate the responsible approach of the US president toward implementing our agreements. I am prepared to continue this dialogue."[20] As it turned out, Obama's missile defense shift in Europe was enough to get Russia's support for New START, but no more than that.

Each of the first three phases of the administration's European missile defense plan came with more capable interceptor missiles. Phase four, however, was in a different league. The SM-3 IIB interceptor, planned for Poland, was intended to defend the United States—not Europe—from an Iranian long-range missile threat that did not exist. Once again, Russia saw the Iran story as a ruse, and assumed the system was meant to intercept Russian ICBMs.

On March 12, 2013, Pentagon policy chief Jim Miller gave a speech on the Obama administration's plans for missile defense in Europe, saying that the first three phases of the system were on track.[21] But, significantly, he did not mention the fourth phase, intended to defend against Iranian ICBMs, which did not exist (and still don't). Then, in response to a question from me (Tom), Miller said, "We are continuing to look very hard at" whether to move forward with phase four or to pursue other options, given budget setbacks and technical issues.[22]

This was shaping up to be a major shift in policy, so I (Tom) wrote an article about it for *Foreign Policy*, which appeared two days later, on March 14. In the article I wrote, "It is time to weed out phase four and let the prospects for US–Russian arms reductions grow."[23] Moscow had been complaining that phase four could be used to intercept Russian ICBMs on their way to the United States.

In the early morning of March 15, I got a phone call I will never forget. I was at the Arms Control Association at the time, and the staffer who answered the main line told me that "the Pentagon wants to talk to you." Feeling like I might be in trouble for something, I nervously picked up the phone. A Department of Defense staffer from the secretary's office was calling to thank me for the article, to congratulate me for being ahead of their policy process, and to make sure I listened in to the secretary of defense's press conference later that day. I said thank you, and I most definitely would.

At the press conference, Secretary of Defense Chuck Hagel said that, under a "restructuring" of the European program, the Pentagon would cancel the last of the four phases of the European Phased Adaptive Approach (EPAA) missile defense system, which would have fielded interceptors in Poland to shoot down any future long-range missiles launched from Iran.

This was welcome news. Until then, the Obama administration had insisted that it would deploy all four phases of EPAA; in December 2010, in order to secure enough Republican votes for the New START Treaty, President Obama explicitly promised the Senate he would proceed with the full plan, assuming the Iranian missile threat continued to develop and the interceptor technology proved effective against it. But the United States did not need phase four, and it had become a significant roadblock to Obama's plans to seek another round of nuclear arms reductions with Russia. It was time to shift gears.

Interestingly, Hagel buried the announcement about canceling phase four under the news that he would add funding to field an additional fourteen ground-based interceptor (GBI) missiles in Alaska by 2017 to address rising nuclear and missile threats from North Korea. Hagel also said that the Alaska system could play a role in countering Iranian long-range missiles, if they appear. He said that by shifting resources "from this lagging program" to the additional GBIs missiles, "we will be able to add protection against missiles from Iran sooner."[24] The Obama administration was covering its tracks on phase four and did not want to be seen as "weak" on missile defense.

Unfortunately, canceling phase four was not enough to bring Moscow back to the negotiating table. Russia was holding out for legally binding limits on US defenses, similar to the ABM Treaty, which the Obama administration was not prepared to offer.

Meanwhile, increasing the Alaska deployment was an irritant to China. "Strengthening anti-missile deployments and military alliances can only deepen antagonism and will be of no help to solving problems," Hong Lei, a spokesman for the Chinese Foreign Ministry, told reporters in Beijing on March 18.

In 2015, in one of its major arms control and nonproliferation achievements, the Obama administration negotiated the Iran nuclear deal (with Russia, China, and the European Union), which verifiably halted any progress Tehran may have been planning to make on nuclear weapons. Unfortunately, the Trump administration abandoned the Iran deal in 2018.

Now Moscow is concerned that missile interceptor sites in Romania and Poland could be used to launch US cruise missiles, which are no longer prohibited by the INF Treaty. Moreover, SM-3 interceptors, like the GMD system, are vulnerable to countermeasures and have not been adequately tested against them.

It is time again to reevaluate US plans for missile defense in Europe. Like phase four, there is no need for phase three in Poland, as Iran's nuclear program has been frozen (although Trump's violation of the nuclear deal threatens to unfreeze it) and there is no reason to believe that Tehran is not deterred by NATO's formidable military forces.

## WHY ANTI-BALLISTIC MISSILES DON'T WORK

The current US midcourse anti-ballistic missile system has been rightly criticized because of its low success rate in tests. But even if all tests *had worked* as planned, the system *still* would not provide an effective defense. Providing an effective defense against ICBMs is extraordinarily difficult. The main problem is not in the design of the interceptors; it is in the laws of physics.

There are three fundamentally different ways to intercept a ballistic missile: during its powered flight into space (boost phase); as it travels through space (midcourse phase); and during reentry into the atmosphere (terminal phase). Each of these approaches has its own problems with the laws of physics.

An ICBM is most vulnerable to attack during its boost phase. But a boost phase defense must have a direct line of sight to the ballistic missile. Unless it is launched near a shoreline, that can only be done from a satellite. Geography might permit the fielding of a seaborne defense system against North Korean ICBMs, but certainly not against those of Russia or China, which have a much greater geographic expanse and base their missiles inland.

The Reagan SDI program was intended to overcome those geographical constraints by basing laser or X-ray beam weapons on satellites that would pass over Soviet launch sites. But this would require many hundreds of satellites to ensure that at least one had a direct shot at each ICBM as it was launched. Alternatively, the beam weapon could be based on a satellite that "hovered" above the launch sites (known as geosynchronous orbits). But those orbits are much higher up in space and would require the beam weapon to be powerful enough to "kill" an ICBM from more than 20,000 miles away. So, the SDI program was never implemented because of the extreme difficulty of overcoming these basic physics problems.

An alternative approach is to attack the warhead during its terminal phase, after it reenters the atmosphere. But there is very little time—just a few minutes—to intercept the warhead before it detonates. Thus, the interceptors must be located near the point of impact to have any chance of success. Such systems (like Patriot) are used in Israel, South Korea, and elsewhere for defending a city or troop deployment from attacks with tactical missiles using conventional warheads. But to defend a large country like the United States from a mass attack of nuclear warheads would require many thousands of interceptors, which would be based in and around metropolitan areas, and even successful intercepts could not prevent many nuclear detonations close to the city being defended (this is sometimes called a "bloody nose" defense).

In the 1950s, a terminal defense approach was briefly pursued to achieve some defense from Soviet bombers, and then abandoned. If you go for a hike in the hills north of San Francisco, you may come across the rusting remains of a Nike Ajax system once briefly deployed there, a stark reminder of the futility of defending a country against a determined air attack. The Soviet Union, on the other hand, did not give up on air defense. It deployed thousands of SA-2 and SA-3 air defense interceptors across the Soviet Union, at a cost of many hundreds of billions of dollars. The US Air Force determined that it could defeat all of those systems by the inexpensive tactic of flying their bombers at 200 feet altitude, underneath Soviet radar beams.

For these reasons, the current US defense system is designed to attack an ICBM warhead during its midcourse phase, when it is in free flight, hundreds of miles above the ground. The good news is that presently deployed US ABMs have line of sight during the free flight of any ICBMs launched from Russia, China, or North Korea. The bad news is that because there is no atmosphere in midcourse, there is no drag, and so the warheads can be simulated by rather simple decoys.

Conceptually, the simplest decoy is just a plastic balloon coated with radar-reflecting material. Such decoys are simple, lightweight, and cheap, and it is hard to imagine that any nation capable of building ICBMs would not have decoys at least that good and have dozens of them deployed on each of their operational ICBMs. Sophisticated radars might be able to discriminate against such simple decoys, so one would expect more capable decoys to be used. For example, it is plausible that warheads could be embedded in a balloon just like the decoy balloon, or hundreds of simple balloons could be deployed, in order to saturate the radar's discrimination capability. In such a competition the advantage is clearly with the attacker, not the defender.

Our present system, the Ground-based Midcourse Defense (GMD), operates during midcourse. There is no reason to have any confidence that this system can defend the country against any plausible ICBM attack. To assert that it can is not only delusional but also dangerous; it can mislead our leaders to take political actions based on the false premise that our country is reliably defended from an ICBM attack.

## SPACE WEAPONS

When Trump released the 2019 Missile Defense Review, he said that the goal of US missile defenses is to "ensure we can detect and destroy any missile launched against the United States anywhere, anytime, anyplace." If the US were to pursue such an unattainable capability, it would likely lead to the deployment of interceptors in space.[25]

As futuristic as they may sound, space-based weapons are an old—and bad—idea. The Reagan administration tried and failed to develop a space-based laser as part of its Strategic Defense Initiative. The George H. W. Bush administration switched from lasers to kinetic kill vehicles with Brilliant Pebbles and, when that failed, came up with Global Protection Against Limited Strikes, or GPALS. They sound almost friendly.

But the false allure of space weapons hides only their immense technical and financial hurdles, not to mention their hugely destabilizing effects. A 2003 American Physical Society study showed that in order to have just one satellite-based interceptor on station above a launch site at any given time would require a network of *at least* 1,600 satellites (with a corresponding five- to tenfold increase in American space-launch capacity).[26] That number nearly matches all the active satellites in orbit today.[27]

Even a bare-bones system would be ridiculously costly. A 2012 National Research Council report determined that the total life-cycle cost of developing, building, launching, and maintaining an "austere and limited-capability network" of 650 satellites would be roughly $300 billion.[28] This is more than the federal government spends annually on the State Department, education, veterans, public housing, and the space program combined.

In addition to spending hundreds of billions for a paper-thin system, space weapons would also spook Russia and China into a dangerous arms race. Since the 1960s, rival powers have maintained a fragile norm against placing weapons in space. The deployment of space-based interceptors would irreparably destroy that precedent. Moreover, an interceptor that is able to target an enemy missile may also be able to knock an enemy satellite out of the sky.[29]

Against this capability, the claim that space-based interceptors have a purely defensive mission would ring hollow in Moscow and Beijing, which would be forced to deploy anti-satellite weapons of their own. This would greatly increase the likelihood of a shooting war in space, posing a grave risk to the satellites upon which the US military (and civil society) depends. As the nation that is most dependent on satellites for military and civil communications, the United States has the most to lose from a space war.[30]

None of this is inevitable, but the development of space weapons greases the skids for this dangerous outcome. The United States should recognize space-based interceptors for what they really are: infeasible, unaffordable, and utterly destabilizing. We should reject space weapons and save our money—and our satellites—instead.

# PART III

# BEYOND THE BOMB

# CHAPTER 9

---

# WHY DO WE STILL HAVE THE BOMB?

*We have lamented our failure to control these weapons at the beginning—when it would, we tell ourselves, have been so much easier.*
—McGeorge Bundy[1]

S eventy-two years after the first atomic bomb was detonated in the New Mexico desert, the United Nations passed the first treaty to prohibit nuclear weapons. On July 7, 2017, 122 states at the UN voted in favor of the Treaty on the Prohibition of Nuclear Weapons, and the pact opened for signature in New York on September 20 of that year, with 50 nations signing on the first day. The treaty will become international law when 50 states have ratified it, possibly some time in 2020.

"This treaty is the beginning of the end for nuclear weapons," said Setsuko Thurlow, one of the last survivors of the atomic bombing of Hiroshima in 1945. "For those of us who have survived the use of nuclear weapons, this treaty gives us hope."[2]

You might ask what took so long. The possession of biological weapons has been banned by treaty since 1972. Chemical weapons have been banned internationally since 1993. Until now, nuclear weapons were the only weapons of mass destruction not yet prohibited by international law, despite the widespread catastrophic consequences that would result from their use.

The ban treaty was the result of a decade-long effort organized by the International Campaign to Abolish Nuclear Weapons (ICAN), based in Geneva. After the treaty was approved, ICAN executive director Beatrice Fihn said: "No one believes that the indiscriminate killing of millions of civilians is acceptable—no matter the circumstance—yet that is what nuclear weapons are designed to do. Today the international community rejected nuclear weapons and set a clear standard against the acceptability of these weapons. It is time for leaders around the world to match their values and words with action by signing and ratifying this treaty."[3]

As of this writing, the treaty has not been signed by any of the nine states that possess nuclear weapons, and so the pact will not eliminate the bomb anytime soon. But the passage of the ban treaty and its growing international support are a clear indication that the majority of the world's nations do not accept nuclear weapons and do not consider them to be legitimate.

Fihn told us in an interview for this book that she believes that the ban treaty, even if the nuclear-armed states don't sign it anytime soon, is "dragging people toward thinking about the elimination of nuclear weapons. And questioning nuclear weapons. And strengthening the notion that there's something wrong with having nuclear weapons. And shaming countries a little bit for having nuclear weapons."[4]

Together, 127 nations have signed a pledge to cooperate "in efforts to stigmatize, prohibit and eliminate nuclear weapons in light of their unacceptable humanitarian consequences and associated risks."[5] These nations know that if there is a nuclear war they will be hurt too, even if they played no role in the conflict and had no say in the outcome. They know that the best way to prevent use of the bomb is to eliminate it from the face of the earth.

———————————————

The ban treaty is the largest and most recent international movement to ban the bomb, but it is not the first. In fact, ever since the bomb was first invented there have been international efforts to ban or control it at the highest levels of government. President Harry Truman, the first and only leader to order the use of atomic weapons in war, soon sought to put them under international control. President Reagan, President Obama, and others pursued similar efforts. Why did they try, and why did they fail?

The elimination of nuclear weapons makes more sense today than ever before, not only for nonnuclear nations but also for the United States, Russia, and the other nuclear states. For the United States, as the nation with the strongest conventional military forces, the nuclear weapons of potential adversaries can act as an asymmetric equalizer for them, undermining US security. Nuclear weapons are also the only immediate existential threat to the United States. Climate change is also a massive threat, but it does not threaten our existence immediately.

"Nuclear war today poses the one existential threat to the United States," wrote Steven Pifer, with the Brookings Institution. "In a nonnuclear world, America would enjoy the advantages of geography (the protection afforded by two wide oceans and friendly neighbors in Canada and Mexico), the world's most powerful conventional forces, and an unrivaled network of allies. Deterrence would not end; US conventional forces could threaten enormous costs to any would-be adversary menacing America or its allies."[6]

Other nations are fully aware that the effects of nuclear war would not be limited to those foolish enough to use nuclear weapons. Radiation and climate impacts would spread throughout the globe, causing great damage to nations that had no say in a decision to unleash the bomb. It is this sense of global danger in the hands of a very few decision makers that gave rise to the ban treaty movement. If we all bear the awful consequences, we should all be part of the solution.

And yet, here we are, on the bomb's 75th birthday, July 16, 2020. The bomb is doing just fine, thank you. It was invented to defeat Hitler, who was dead before it was first tested. It was used to defeat Imperialist Japan, now replaced with a democratic nation with no imperialistic ambitions. It multiplied to deter the Soviet Union and "win" the Cold War, both of which ended three decades ago. Most Americans have long forgotten about mushroom clouds, fallout shelters, and "duck and cover," yet the bomb remains.

Despite successful efforts to cut arsenals, the bomb has been normalized. It is just one more thing we wish we could do something about but cannot, or so we think. The numbers of weapons in the world have declined dramatically but still remain alarmingly high, yet efforts to reduce those numbers further appear to have stalled. Like a mirage in the desert, the closer we get to zero the harder it is to get there. The countries that have the bomb may talk about elimination, but their actions indicate that they are determined to keep it. In fact, Russia and the United States, the two countries that have about 90 percent of the world's nuclear weapons, are both in the early stages of building a new generation of bombs to replace the ones now deployed.

Ultimately, the world cannot escape the dangers of nuclear war as long as the weapons remain. Gorbachev wrote in 2019 that "nuclear deterrence, instead of protecting the world, is keeping it in constant jeopardy."[7] Given what we know about the dangers of atomic weapons, can we survive another 75 years? Let's take a close look at past efforts to control the bomb, why they failed, and how we might try again.

## BEST CHANCE LOST: INTERNATIONAL CONTROL

Even before the United States dropped the bomb on Japan, an enlightened few were thinking about what would come next. In one of the earliest examples of atomic scientists seeking to influence political leaders, Chicago physicist and Nobel laureate James Franck formed a group to consider the future of the bomb. He teamed up with Eugene Rabinowitch, who later founded the *Bulletin of the Atomic Scientists*, and Leo Szilard, who in 1939 wrote the letter for Albert Einstein's signature to President Roosevelt that resulted in the

Manhattan Project. They argued that instead of dropping the bomb on Japan without warning, it would be preferable to stage a demonstration, and that this path might improve the possibilities for postwar international control of atomic weapons:

> It will be very difficult to persuade the world that a nation which was capable of secretly preparing and suddenly releasing a weapon, as indiscriminate as the rocket bomb and a thousand times more destructive, is to be trusted in its proclaimed desire of having such weapons abolished by international agreement.[8]

But military considerations, such as maintaining the element of surprise in an attack, held sway and the United States bombed Hiroshima without warning.

The Franck report also argued that if the first use of the bomb was not followed by an international agreement to control it, there would be a "flying start of an unlimited armaments race."[9]

Danish physicist Niels Bohr also saw an opportunity for international cooperation on nuclear energy and the dangers of a possible arms race. Bohr argued that before the bomb was used, the United States and Great Britain should discuss it with the Soviet Union.[10]

Bohr met with British prime minister Winston Churchill in May 1944, but Churchill did not agree with Bohr. Bohr then met with Roosevelt, who did agree that he and Churchill should talk to Stalin. Churchill remained opposed, but in any case, Roosevelt died before the subject could be broached with the Russians at Potsdam, Germany, in July 1945.

The scientists continued to press their case, suggesting that "Russia, France, and China be advised that we have made considerable progress in our work on atomic weapons, that these may be ready to use during the present war, and that we would welcome suggestions as to how we can cooperate in making this development contribute to improved international relations."[11]

This early effort was blocked by US and British distrust of the Soviets. The scientists argued, correctly, that Russia could make its own bomb in three or four years, and Churchill, in particular, was concerned that giving the

Soviets too much information could accelerate their program and erode the US–British advantage. This paranoia would come back to haunt them.

By the time Truman got to Potsdam, his mission was to tell Stalin the bare minimum, which is what he did. It was nine days after the first bomb test at Alamogordo, New Mexico, and twelve days before Hiroshima, and the United States had never had a conversation about the bomb with its most powerful wartime ally and most likely postwar competitor. As Truman recorded in a letter, their eventual exchange was brief:

> TRUMAN: We have a new weapon of unusual destructive force.
> STALIN: Glad to hear it, I hope you'll make good use of it against the Japanese.[12]

And that was it. Truman told Stalin of a new powerful weapon but did not even mention it was nuclear. And he made no effort to start a dialogue on what would become the nuclear age.

Historian David Holloway argues that even if Truman had told Stalin about the bomb at Potsdam, Russia would have built it anyway. "As the most powerful symbol of American economic and technological might, the atomic bomb was ipso facto something the Soviet Union had to do," Holloway said.[13]

In any case, the sad irony is that Stalin already knew about the Manhattan Project from his spies in the program and had already begun serious atomic research under Russian scientist Igor Kurchatov. By not sharing information that Russia already had, Truman only succeeded in confirming Stalin's fears of American deceit.

As an advisor to Churchill wrote to Truman, "While it may be wise to keep the secret to ourselves for the moment, it will not remain a secret long, and its disclosure after the war may start the most destructive competition in the world.... If ever there was a matter for international control, this is one."[14]

We will never know whether the postwar effort to achieve international control of nuclear energy would have fared better if this wartime outreach to Stalin had been handled in a way that built trust, rather than undermined it. But as Bundy said, the way it was handled "may well have made a hard prospect even harder."[15] It should have been clear to Truman and Churchill that the key to international control was Moscow. If any nation was going to get the bomb second, it was Russia.

Truman authorized the bomb to be dropped on Japan, and everything changed. He wrote in his diary, "We must constitute ourselves trustees of this new force—to prevent its misuse, and to turn it into the channels of service to mankind."[16] In this, he failed miserably. By the end of his time in office, the United States would have about a thousand nuclear weapons and the arms race would be in full swing.

The war with Japan now over, Secretary of War Henry Stimson began to see the wisdom of engaging Russia on the bomb. Before he retired, Stimson urged Truman to make a proposal to the Soviets. At his last cabinet meeting on September 21, 1945, Stimson made his case for direct talks with Russia. It turned into a long and heated discussion that many saw as a dangerous plan to share bomb secrets with Moscow. What happened next shows how little has changed in Washington in the past seventy-five years.

The Army and Navy were opposed to sharing nuclear information with the Soviet Union, and to defeat such sharing, they leaked their version of the issue to the *New York Times*, which ran this headline: "Plea to Give Soviet Atom Bomb Secret Stirs Debate in Cabinet . . . Armed Forces Opposed." The story said that Stimson was urging talks with Russia at the United Nations, but that the Army and the Navy "are prepared to resist the proposal to the hilt."[17]

Thus began a long tradition of military opposition to nuclear diplomacy. Hardly an auspicious start.

The discussion continued inside the Truman administration. Under Secretary of State Dean Acheson argued that "if the invention is developed and used destructively there will be no victor and there may be no civilization remaining." He warned, "The joint development of this discovery with the UK and Canada must appear to the Soviet Union to be unanswerable evidence of an Anglo-American combination against them." He concluded, "For us to declare ourselves trustee of the development for the benefit of the world will mean nothing more to the Russian mind than an outright policy of exclusion."[18]

Acheson called for an immediate opening to the Russians, discussions with the British and the US Congress, and "informed and extensive public discussion." Without such discussion, he warned that "the public and Congress will be unprepared to accept a policy involving substantial disclosure to the Soviet Union."

That the Congress was unprepared to accept a sharing policy was soon manifest. Senator Tom Connolly of Texas, Democratic chairman of the Senate Foreign Relations Committee, said on the day the *Times* story broke that "complete secrecy should be maintained regarding the atomic bomb." It is typical for such debates, then and now, to quickly become polarized and exaggerated. In reality, an opening to Russia on the bomb did not have to reveal details beyond what had already been disclosed, at least initially. Acheson remarked that such a dialogue with Moscow "need not involve at this time any disclosures going substantially beyond those which have already been made to the world."

President Truman sent his plan to control the bomb to Congress on October 3, 1945, just two months after the bombs were dropped. Truman declared:

> Civilization demands that we shall reach at the earliest possible date a
> satisfactory arrangement for the control of this discovery in order that it
> may become a powerful and forceful influence toward the maintenance
> of world peace instead of an instrument of destruction. . . . The hope
> of civilization lies in international arrangements looking, if possible,
> to the renunciation of the use and development of the atomic bomb,
> and directing and encouraging the use of atomic energy and all future
> scientific information toward peaceful and humanitarian ends. The dif-
> ficulties in working out such arrangements are great. The alternative to
> overcoming these difficulties, however, may be a desperate armament
> race which might well end in disaster. . . .
>
> Discussion of the international problem cannot be safely delayed
> until the United Nations Organization is functioning and in a position
> to adequately deal with it. I therefore propose to initiate discussions,
> first with our associates in this discovery, Great Britain and Canada,
> and then with other nations, in an effort to effect agreement on the

conditions under which cooperation might replace rivalry in the field of atomic power."[19]

Despite this fine rhetoric, Truman apparently had no intention of sharing the secrets of the bomb with Russia. As he later said, "If they catch up with us on that, they will have to do that on their own hook, just as we did."[20] Truman, in agreement with his military advisors, saw no conflict between secrecy about the bomb and seeking international control.

Truman's UN plan was informed by the March 1946 Acheson-Lilienthal report, which called for an international "Atomic Development Authority" that would have a global monopoly on control of nuclear materials, from mining to manufacturing, from uranium to plutonium, from enrichment plants to nuclear reactors (beyond laboratory size). The organization would seek to control the bomb as well as promote nuclear energy. The logic was that banning the bomb was impractical, because it would require intrusive inspections that would not be possible. But if the world's raw materials for the bomb were controlled by an international organization, no single nation would be able to build a bomb in secret.

The report also proposed that the United States give up its monopoly on the bomb and offer to tell the Soviets key information if it agreed not to build its own atomic weapons. But Washington planned to keep its bombs until the details of international control could be worked out.

## THE BARUCH PLAN

At this time, the UN was still brand new, having been founded on October 24, 1945, in San Francisco. Nuclear disarmament is the United Nations General Assembly's oldest aspiration. Its first resolution, adopted on January 24, 1946, urged the "elimination from national armaments of atomic weapons and of all other major weapons adaptable to mass destruction."

President Truman gave Bernard Baruch the job of bringing his plan for international control of the atom to the UN. Secretary of State James Byrnes, who recommended Baruch to Truman, later admitted that the choice was "the worst mistake I ever made."[21] An elder statesman beyond his prime, Baruch was chosen more for his ability to reassure the US Senate and the public than to work effectively with the Soviets.

Baruch introduced the plan at the UN in June 1946, and by the end of the year it was dead. Instead of international control, the Soviets said they wanted a prohibition on the possession, production, and use of atomic weapons, and they rejected a key part of the Baruch plan—that, in the case of a violation, no state would be able to veto a UN decision on how to respond, including war. The Soviets saw this as a direct challenge to their veto power as a permanent member of the UN Security Council.

Baruch pushed for a Security Council vote in December, and although it passed 10-0, Russia and Poland abstained, and it was clear the issue could not move forward without Soviet support. Baruch resigned in January 1947. Talks continued at the UN in 1947 and 1948, but during this time, there were no direct behind-the-scenes talks between Washington and Moscow to narrow their differences. Soviet opposition only hardened. Moscow would accept no limitations on its atomic program. And we now know why.

Soon after Hiroshima, in mid-August 1945, Stalin decided to go for the bomb too. "A single demand of you, comrades," Stalin said to his scientists, "provide us with atomic weapons in the shortest possible time. . . . Provide the bomb—it will remove a great danger to us."[22]

To the Soviets, Baruch's demand to give up the option for the bomb in exchange for an international organization with a monopoly on the bomb meant giving up their top-secret program to neutralize the US advantage and secure their future. The American offer was seen as a ploy to maintain its advantage. Given the lack of trust between the two nations, and Stalin's fear that Russia could be the next target, his decision to go nuclear was simple and swift.

Nor did the Americans trust the Soviets' diplomatic offer. Advisor George Kennan wrote in 1946 that the Russian proposal to ban the bomb was intended to "effect the earliest possible disarming of the United States with respect to atomic weapons."[23]

As Bundy notes, this moment was likely humanity's best chance to control the bomb and prevent the arms race and Cold War to come. To miss it was a failure of political imagination, to fully grasp what was next and the value of preventing it.

The Baruch plan may have been doomed to fail, but what about other approaches? Even if we accept that Stalin would never let the United States keep its monopoly on the bomb, what if they had agreed to limits on numbers and sophistication? Achieve parity at a low level, then stop, and agree to ban the hydrogen bomb? The United States and the Soviet Union eventually saw the wisdom of agreed numerical and testing limits, but not until they wasted blood and treasure on tens of thousands of weapons of increasing sophistication.

The effort for international control failed for reasons that should come as no surprise: the US distrust of the Soviets, and the Soviet distrust of the United States; an early addiction to secrecy, a reluctance to share secrets, and a fear that such sharing would help the Soviets get the bomb; the emerging politics of the bomb, where politicians wanted to look "tough" and were afraid of appearing "weak" in the eyes of the public; military opposition to cooperative approaches to security; the newness of atomic weapons and the lack of experience in how to control them; the leaking of secret information to serve a political agenda; the realization that people are policy—if you don't have the right messenger, the message might not matter; Roosevelt's sudden death and Truman's newness in his role, where a new post-war nuclear order was not being pushed from the top down; and finally, the lack of imagination on all sides as to how daunting the consequences of failure would be.

Truman and Stalin's failure to control the bomb at the very start meant that the atomic weed would now grow and firmly establish itself. In the next decade, the number of weapons would rise, the budgets would increase, and the numbers of people and jobs involved would skyrocket. Future efforts to control the bomb would now have to take on a formidable new force—the ever-expanding nuclear bureaucracy that had a strong vested interest in building and deploying more bombs.

## THE "SUPER"

The Soviets tested their first nuclear device in August 1949, and almost imme-diately there were efforts in the United States to move ahead with a weapon of even greater magnitude. The US answer was the hydrogen bomb, known as the "super." It had been known to a small circle of insiders in the United States since 1942 that thermonuclear weapons were theoretically possible, taking their energy primarily from fusion rather than fission, and producing explosive power a thousand times greater than the Hiroshima bomb.

"There is an immense gulf between the atomic and the hydrogen bomb," Winston Churchill said in 1955, and with the new weapon "the entire foun-dation of human affairs was revolutionized, and mankind placed in a situa-tion both measureless and laden with doom."[24]

The importance of the decision on the super was second only to Roo-sevelt's order to build the fission bomb in 1941. The General Advisory Com-mittee of the Atomic Energy Commission met to consider the issue in October 1949. The committee was chaired by none other than Robert Oppenheimer, technical director of the Manhattan Project, and included science luminaries Enrico Fermi, I. I. Rabi, and Glenn Seaborg, among others.

At the end of a two-day meeting, the eight members reached a unani-mous conclusion:

> We all hope that by one means or another, the development of these weapons can be avoided. We are all reluctant to see the United States take the initiative in precipitating this development. We are all agreed that it would be wrong at the present moment to commit ourselves to an all-out effort toward its development.[25]

In addition, two members, Fermi and Rabi, thought that this unquali-fied commitment should be conditioned on "the response of the Soviet gov-ernment to a proposal to renounce such development." The majority of the group found that "the extreme dangers to mankind inherent in the proposal wholly outweigh any military advantage that could come from this develop-ment . . . a super bomb might become a weapon of genocide."

The General Advisory Committee delivered its unanimous finding to the AEC, which then split on the issue, three to two. Lilienthal, the chairman, was opposed to the super, and he reported to Truman his view that the existing nuclear arsenal already was an adequate deterrent against Russia. Meanwhile, Congress weighed in. Democrats were under pressure for being "soft on Communism," and Senator Brien McMahon, chairman of the Joint Committee on Atomic Energy, argued for the super and against the scientists' report. The senator said that "if we let Russia get the super first, catastrophe becomes all but certain."[26] Gen. Omar Bradley, chairman of the Joint Chiefs of Staff, wrote it would be "intolerable" to let the Soviets get the thermonuclear bomb first, and that a US "unilateral" decision not to develop the bomb would not stop others from doing so.

Truman met with his national security team and asked if the Russians could build a super, and all agreed that yes, they could. "In that case," Truman said, "we have no choice. We'll go ahead."[27] The meeting lasted seven minutes. Lilienthal wrote later that Truman was "clearly set on what he was going to do before we set foot inside the door." The Joint Chiefs had won. The next day's *New York Times* read, "Truman Orders Hydrogen Bomb Built." The rapid development of the super became US policy, and Truman apparently never considered the issue again.

Soon after Truman's decision, Albert Einstein read a prepared statement on national television. He criticized the "hysterical character" of the nuclear arms race and said the decision to build the super was based on a "disastrous illusion." "Every step appears as the unavoidable consequence of the preceding one," he said. "In the end, there beckons more and more clearly general annihilation."[28]

The first US hydrogen bomb test occurred on November 1, 1952, on the island of Elugelab in Enewetak Atoll, in the Pacific Ocean. Its yield was 10.4 megatons (10,400 kilotons) of TNT, or five hundred times larger than the Hiroshima bomb. Russia tested its first fusion device less than one year later on August 12, 1953. The largest H-bomb test ever was the Soviet Union's Tsar Bomba in 1961, with a yield of 50 megatons. (The Tsar Bomba actually had a yield of 100 megatons, but the scientists scaled it down because they feared the delivery aircraft would not be able to get far enough away to avoid damage from the blast. That fear turned out to be justified, because even at the reduced yield and with the bomb being dropped by parachute to allow the aircraft to get farther away from the detonation, the aircraft received considerable blast damage.)

China tested its first thermonuclear bomb in 1967. Fifty years later, in 2017, North Korea surprised the world by testing its own thermonuclear weapon. All the early discussions about whether to build H-bombs were based on a consideration of actions by the Soviet Union. The possibility that a small rogue nation like North Korea could build such a weapon was never considered.

It seems that Truman never seriously considered the alternative (if he even knew of it) proposed by Fermi and Rabi to "invite the nations of the world to join us in a solemn pledge not to proceed in the development or construction of weapons in this category."[29] It was clear that the main audience for such a proposal would be Russia, and that this would essentially be an agreement not to test thermonuclear weapons, which could have been verified by national means of remote detection. (High-yield thermonuclear tests would be readily detectable without intrusive inspections inside Russia.) But this idea was never pursued.

By this time, the dream of international control of the bomb was dead and US–Soviet mutual distrust was alive and well. The military's "full steam ahead" attitude about the bomb clearly held sway with Truman and others over the scientists' calls for bilateral cooperation. The Cold War was now in full swing, and the Soviets were the enemy. As Bundy put it, "Truman and Acheson had learned not to trust the Russians, and both of them now had more critics on the right than on the left."[30]

It is notable that the majority of the General Advisory Committee opposed the super regardless of what the Soviets might do. They did not want to be forced into supporting the H-bomb if a diplomatic opening to Russia failed, as it likely would have (though this possibility should have been tested). To them, the super was simply not the right answer to the Soviet fission test. But the political pressure to go ahead was powerful.

As it turns out, the Soviets had been working on a hydrogen bomb since 1948. According to Soviet physicist Andrei Sakharov, Stalin was committed to building an H-bomb regardless of what America did. "Any US move toward abandoning or suspending work on a thermonuclear weapon would have been perceived either as a cunning, deceitful maneuver or as evidence of stupidity or weakness," Sakharov wrote in his memoirs. "In any case, the Soviet reaction would have been the same: to avoid a possible trap and to exploit the adversary's folly."[31]

Thus began a period of increasing political polarization about the bomb. Some opposed the ongoing development of nuclear weapons, while others opposed any efforts to limit that development. There were few people supporting compromise solutions, such as continuing development of the super but withholding a decision to test and build the weapons. Truman could even have continued development of the new weapon while pursuing a thermonuclear test ban with Russia.

Another notable feature of this debate, which we will see again and again, is that power was growing among the advocates for the bomb, and those opposed were losing influence. For Truman to oppose the super, or even slow it down, would have been more difficult and politically risky than the path he chose. To oppose the super would have required Truman to go against the military and powerful forces in Congress, and to open a public debate in which he would have to clearly lay out his case for a world without the H-bomb. He might have won this debate, but it would have taken valuable time and effort, and he did not try. He took the easy way—and the easy way was, and still is today, to build more nuclear weapons, not fewer.

Truman went on to approve eight plutonium production reactors and two uranium enrichment plants to support a massive expansion of the nuclear arsenal, which grew from 1,000 bombs in 1953 to nearly 18,000 by 1960, peaking at 31,000 in 1967.[32]

Despite Truman's example here, presidents do *not* always take the easy way. As Truman did by taking sole authority for the bomb, they can choose to challenge conventional wisdom, stand up to the nuclear bureaucracy, take political risks, and try to do the right thing. It is in these times that true leadership emerges. For when it comes to progress on nuclear disarmament, there is one key ingredient that is required for success: presidential leadership. And given how strong the current is flowing for the bomb, it is notable and surprising when leaders decide to swim upstream.

## THE RISE AND FALL OF THE ABM TREATY

By the late 1960s, Soviet missile forces started to catch up to earlier, exaggerated predictions. Russian ICBMs increased to a thousand by 1969, and they were larger if less sophisticated than the US Minuteman. At the same time, anti-missile systems were beginning to mature. The Johnson administration had "irrefutable evidence" that the Soviets were deploying an anti-ballistic missile (ABM) system around Moscow to defend the city against long-range US missiles. The administration assumed—incorrectly—that the Soviets planned to field the system across their vast nation. As McNamara argued, "Why would anyone put a system around one city and nowhere else?"[33]

At first, Congress supported a US response in kind. The Army had a system called Nike Zeus and then Nike X, which Congress funded in 1966. But the Johnson administration refused to build it because the president and McNamara believed that it was ineffective and would provide little, if any, protection.

Nevertheless, the Joint Chiefs of Staff were pushing for ABMs. In a December 1966 meeting with President Johnson and the chiefs, McNamara was in a weak position—Congress had funded the program and the Joint Chiefs wanted it. So McNamara proposed that the administration support the ABM program but hold off on a deployment decision until "after we make every possible effort to negotiate an agreement with the Soviets which will prohibit deployment of defenses by either side and will limit offensive forces as well."[34]

The president recognized a good compromise and grabbed it.

In June 1967, at the height of his struggles with the Vietnam War that would hobble his presidency, Johnson met with the Soviet premier, Alexei Kosygin, in Glassboro, New Jersey, to discuss ABMs. McNamara explained that deployment of ABMs would lead to an escalation of the arms race, and "that's not good for either one of us." Kosygin banged the table and said, "Defense is moral; offense is immoral!"[35] End of discussion.

In the face of what appeared to be Soviet determination to deploy ABMs, the Johnson administration decided to respond by expanding its offensive forces. The cheapest way to do that was to place many warheads (each of which could be aimed at a separate target) on each missile, called multiple independently targetable reentry vehicles, or MIRVs. But this was also very dangerous. If the Soviets followed the US lead, both sides would dramatically increase their offensive forces. So the administration decided to develop MIRVs but not deploy them until it could explore an agreement to ban defenses. According to McNamara, "If such a treaty was negotiated, the MIRV program would be scrapped."[36]

We got the treaty to ban nationwide defenses but, in yet another major missed opportunity, the MIRV program remained.

The 1968 elections brought Richard Nixon into the White House, and the new administration continued Johnson's pursuit of arms control with the Soviets for three reasons: to preserve rough parity of US and Soviet strategic weapons; to save money by maintaining parity at lower levels and forgoing expensive new weapons; and to reduce uncertainty in their relationship, "making both sides less nervous about potential threats to its strategic capabilities," in Kissinger's words.[37]

Like Johnson, President Nixon supported anti-missile interceptors but, surprisingly, these proposals had lost popularity in Congress. The scientific community was skeptical, given the technical difficulty of missile defense. The administration itself thought the Pentagon's anti-missile plans were of low value, finding that deploying interceptors to protect Minuteman ICBMs at four sites "will save only about ten Minutemen more than no defense at all."[38] News of possible missile deployments sparked local opposition—no one wanted to be a target. And there was the larger strategic point—defenses were bound to provoke bigger and more sophisticated offenses. Vice President Spiro Agnew had to break a tie vote in Congress (where the Democrats controlled both houses) to pass Nixon's anti-missile program.

Seeing the political writing on the wall, Nixon decided to make the best of a weak hand and use his anti-missile program as a bargaining chip with Russia. For the Soviets, who feared America's technical advantage, General Secretary Leonid Brezhnev and a new generation of leaders sought to promote détente through arms control and were determined to forge a more peaceful relationship with Washington.

As part of the Strategic Arms Limitation Talks Agreement (SALT I) in 1972, both sides agreed to limit themselves to one site for a local ABM defense and to ban nationwide defenses. This was a huge victory for nuclear sanity on two levels. First, it held back a new weapons system, anti-missile interceptors, that threatened to open a new avenue in the arms race. And second, those interceptors would have stimulated an arms race for offensive weapons to overcome the defenses.

But it would be wrong to see the ABM Treaty as an example of inspired leadership by Nixon or Kissinger, his national security advisor. They had wanted to deploy anti-missile systems, and turned to strike a deal with Russia only when they did not think they could win ongoing support in Congress.

Moreover, although the SALT I agreement placed limits on long-range missiles, it failed to control MIRVs. Indeed, it actually provided a perverse incentive to build more strategic warheads, which in the United States increased from 1,800 in 1970 to 6,100 in 1975. The Soviets, who were lagging in MIRV technology, expanded from 1,600 to 2,500.

Why did Nixon and Kissinger leave MIRVs uncontrolled? As William Hyland, a close confidant of Kissinger, wrote in *Mortal Rivals* (1987),

> Refusal to ban MIRVs was the key decision in the entire history of SALT I. Both Nixon and Kissinger thought it would be a weak move at the outset of a new administration and the opening of a long negotiation. And it would have provoked a bloody fight inside the administration and in the Congress. It was a truly fateful decision that changed strategic relations, and changed them to the detriment of American security.[39]

Nixon and Kissinger had set out to deploy missile interceptors and, thanks to congressional opposition and smart diplomacy by chief US negotiator Gerard C. Smith and others, got the 1972 ABM Treaty instead. But they failed to mount the effort it would have taken to ban MIRVs. As Kissinger said in 1974: "I would say in retrospect that I wish I had thought through the implications of a MIRVed world more thoughtfully in 1969 and 1970 than I did."[40]

(Ironically, twenty years later, President George H. W. Bush signed the

START II Treaty with Russia that finally banned MIRVs. But that treaty never went into force, in part because Russia rejected it after President George W. Bush withdrew from the ABM Treaty in 2002. Soon thereafter, the Russians began building a new class of MIRVed missiles.)

Despite the failure to control MIRVs, the ABM Treaty was a success and created the foundation for future arms reductions. But like arms control itself, the treaty was based on a counterintuitive idea: that it is better to leave yourself vulnerable to nuclear attack than to try to defend yourself. It was only a matter of time before a politician would come along and challenge that notion.

## REAGAN'S SHIELD OF DREAMS

President Reagan was never comfortable with the idea of mutual vulnerability, that Americans could be attacked at any time by Soviet nuclear missiles, no matter how unlikely that might be. So, he proposed a dual solution: build a missile defense system that would block a Soviet attack, and at the same time seek to eliminate nuclear weapons through diplomacy. It seems that Reagan did not understand the incompatible nature of these goals. Ultimately, Reagan's commitment to the first was stronger than to the second.

Messages were mixed from the start. Reagan launched his Strategic Defense Initiative (SDI), known popularly as "Star Wars," in March 1983. In his January 1985 inaugural address, Reagan said, "We seek the total elimination one day of nuclear weapons from the face of the earth."

Reagan told high school students in Glassboro, New Jersey, in 1986—the same spot where Johnson met with Soviet premier Kosygin to talk missile defense—that his ideal missile defense would "enable us to put in space a shield that missiles could not penetrate, a shield that could protect us from nuclear missiles just as a roof protects a family from rain."[41] This goal is as appealing as it is technically unattainable.

Progress on Reagan's "Star Wars" dream never got far enough to cause the administration to withdraw from the ABM Treaty, but it was clear that Reagan would do so if that was necessary for his space shield to become reality.

The money and time spent by the Reagan administration on SDI might have all been worth it had Reagan, like Nixon, been willing to trade missile defense away in exchange for a diplomatic deal with the Soviets. But it was not to be.

In October 1986 at the Reykjavik Summit with Gorbachev, Reagan flatly refused to agree to limit the testing of long-range defensive systems. This was a historic missed opportunity to eliminate nuclear weapons, second only to Truman and Stalin's failure to achieve international control in 1946.

At the summit, Reagan asked Gorbachev if he would support a proposal such that "all nuclear explosive devices would be eliminated, including bombs, battlefield systems, cruise missiles, submarine weapons, intermediate-range systems, and so on."[42]

Gorbachev agreed.

"Then let's do it," said George Shultz, Reagan's secretary of state.

This most far-reaching proposal in arms control history was undone by Reagan's misplaced belief in a national missile defense system—which more than thirty years later is no closer to being realized. Gorbachev wanted missile defense research to be limited to laboratory experiments, and Reagan refused. Reagan missed a golden opportunity to make the world a much safer place.

The two leaders made it clear to all that neither nation had any intention of attacking the other with nuclear weapons. In fact, they had been open to eliminating all of them. This diplomatic momentum was strong enough to achieve the Intermediate-Range Nuclear Forces (INF) Treaty in 1987, and then in 1991 President George H. W. Bush and Gorbachev signed the START Treaty.

But Reagan had also solidified the hope and possibility of missile defense in the American imagination. From now on, it would be much more difficult for politicians to resist the impulse to support missile interceptors to "defend America" from nuclear attack. Reagan had made missile defense respectable—indeed, popular—but his plan was still too big, expensive, and complex. It was, in fact, unachievable.

It would take President George H. W. Bush and his son George W. Bush to bring missile defense down to earth. They talked about missile defenses aimed not at a full-scale Russian attack, which most experts saw as impossible, but at an accidental launch of a few missiles from Russia or China or an attack from a rogue state like North Korea. And when North Korea conducted its first satellite launch in 1998, the stage was set for President George W. Bush to withdraw from the ABM Treaty in 2002 and field a rudimentary system to achieve that limited objective.

As we saw in the previous chapter, now that the United States has fielded a long-range anti-missile system, it is hard to imagine we will ever get rid of it, no matter how ineffective it is. Thus, it is hard to see how Washington could ever recreate a legally binding limit on missile interceptors, as was embodied by the ABM Treaty. Yet this is what Russia sees as a precondition to move ahead with nuclear arms reductions.

This is a major reason why the bomb is still with us. The Republican campaign to deploy missile defenses, started by Reagan and delivered by George W. Bush, has essentially created a floor below which Russia will not reduce its strategic nuclear weapons. Unless we somehow remove that barrier, US–Russian arms reductions will only go so far.

## FUMBLING THE END OF THE COLD WAR

The end of the Cold War and the collapse of the Soviet Union brought a rare opportunity for the United States to not only seek to reduce nuclear weapons, but also to transform its relationship with Russia from antagonism to something better. Simply put, we blew it. Thirty years later, the US–Russian relationship is at an all-time low.

The decade after the Soviet collapse was difficult for most Russians. They were going through a deep economic recession, crime was rampant, President Boris Yeltsin was an embarrassment, and they felt disrespected by other nations, particularly the United States. Russians blamed their problems on Gorbachev, on their new democracy, and on Washington, which Russians thought was taking advantage of their weakness to keep them down. In a troubling sign, some Russians began to yearn for the "good old days" of the Soviet Empire, before the collapse of the repressive Soviet system.

There was an historic opportunity here for the NATO alliance, which had been established to keep the Soviet Union out of western Europe, to open up to Russia and forge a new partnership.

As the Soviet-led Warsaw Pact dissolved, eastern European nations sought to join NATO. However, NATO had no plan for how to do this without alienating Russia. As much as NATO wanted to welcome these new nations, if not handled correctly this process could scuttle a historic opportunity to work cooperatively with Moscow to reduce the threat from nuclear weapons.

President George H. W. Bush's secretary of state James Baker assured Gorbachev that NATO would expand "not one inch eastward" during a February 9, 1990, meeting, and documents show that Gorbachev only accepted German reunification—over which the Soviet Union had a legal right to veto under treaty—because he was told that NATO would not expand after he withdrew his forces from eastern Europe.[43] Gorbachev received such assurances from James Baker, President George H. W. Bush, West German foreign minister Hans-Dietrich Genscher, West German chancellor Helmut Kohl, CIA director Robert Gates, French president François Mitterrand, British prime minister Margaret Thatcher, British foreign minister Douglas Hurd, British prime minister John Major, and NATO secretary-general Manfred Woerner.[44]

The United States was not prepared to offer NATO membership to Russia, but instead proposed to set up the Partnership for Peace (PFP), into which former Warsaw Pact states would be invited to join. This would be an auxiliary of NATO and stop short of membership, but it could be a step toward that end. Russia would be invited to join the PFP.

To help build a new partnership with Russia, I (Bill) met with Russian defense minister Pavel Grachev at NATO in 1993, when I sat in for Defense Secretary Les Aspin (I then served as his deputy). After succeeding Aspin as secretary of defense in 1994, I hosted a dinner in Grachev's honor with all NATO defense ministers. Grachev was invited to the NATO meeting that year as a way of establishing good relations between Russia and NATO.

A period of remarkable collaboration followed. Russia agreed to join the PFP, Grachev was allowed to attend NATO meetings (without voting rights), and Moscow appointed a senior officer to serve as a permanent representative. Grachev chose a first-class officer for the post, who later told me: "I spent most of my career doing detailed planning for a nuclear strike on NATO forces. I never dreamed that I would be standing here at NATO headquarters, talking with NATO officers, and planning joint peacekeeping exercises!"[45] Russia participated in PFP exercises in the United States and Ukraine, and hosted PFP exercises that included troops from the United States, other NATO nations, and Ukraine. These peacekeeping exercises turned out to be valuable training for the peacekeeping operation soon to take place in Bosnia.

When NATO deployed military forces into Bosnia in late 1995, Russia offered to send one of its best paratrooper brigades to join the effort, which was led by an American general. To work this out, I met with Grachev in the United States, who at first insisted that his troops could not report to a NATO commander. So, I took Grachev on a tour of Fort Riley, where he rode one of our ceremonial cavalry horses, and then to Whiteman Air Force Base, where he sat in the pilot seat of a B-2 bomber. All the while we were discussing how to resolve the thorny question of how to involve Russian troops in the NATO operation about to take place in Bosnia. Grachev finally agreed to let his troops serve under the "tactical command" of an American commander, as long as it was not a NATO commander.[46]

The success of the Bosnia mission showed how effective NATO could be with European-wide cooperation, including Russia. There was an effort to create some version of the Marshall Plan for eastern European nations with desperate economies and fledgling democracies. Unfortunately, this did not materialize.

But the PFP had proven its value, and now its members wanted to join NATO. Yet Russia still saw NATO as a potential threat, even more so if its former buffer states were to become members. In my estimation, the time was not yet right to push for NATO enlargement. We needed more time to bring Russia into the Western security circle. I (Bill) wanted to delay the process for a few years so that Moscow had more time to work with NATO and would not react negatively to new states joining.

Then, in 1996, fresh off his success at the Dayton Peace Accords, State Department assistant secretary Richard Holbrooke proposed that NATO invite Poland, Hungary, the Czech Republic, and the Baltic states to join right away. I disagreed and went to President Clinton and explained my concerns. The president called a meeting of the National Security Council, and I made my case. Surprisingly, neither Secretary of State Warren Christopher nor National Security Advisor Anthony Lake spoke out. Vice President Al Gore argued in favor of immediate expansion. Clinton agreed with Gore, and he approved immediate membership for Poland, Hungary, and the Czech Republic, but delayed consideration of the Baltic states until later. Clinton and Gore believed we could manage the problems with Russia. It turned out they were incorrect.

This pivotal meeting was the beginning of the end of our warming relationship with Moscow. Looking back, I regret that I did not fight more effectively for my position. Prior to the NSC meeting, I could have met with Christopher and Lake to rally them to my side. I could have written a paper laying out my case to be distributed before the meeting. Afterward, I considered resigning but decided that I would stay and try to help reduce the growing mistrust.

In an open letter to President Clinton, more than forty foreign policy experts, including Bill Bradley, Sam Nunn, Gary Hart, Paul Nitze, and Robert McNamara, expressed their concerns about NATO expansion as both expensive and unnecessary given the lack of a Russian threat at that time.[47]

As George Kennan said in 1998, "I think [NATO expansion] is the beginning of a new Cold War. I think the Russians will gradually react quite adversely and it will affect their policies. I think it is a tragic mistake."[48]

From then on, NATO kept Russia at arm's length and expanded its membership to include former Soviet states. Moscow saw this expansion as a threat and regarded the 2004 inclusion of the Baltic states (which had been part of the Soviet Union for decades) as "marching the NATO threat up to their border." With a tragic lack of forethought, the United States and NATO essentially acted as if Moscow's concerns did not matter.

Russia was particularly alarmed by NATO's actions in Kosovo (which were carried out without UN or Russian approval), the fielding of missile interceptors in eastern Europe, and the continued march of NATO expansion to possibly include Georgia and Ukraine.

When President Obama came into office in 2009, he announced he would try to repair the damage and seek to "press the reset button" on

US–Russian relations. For a while it seemed to work, and President Dmitry Medvedev (who took over temporarily from Putin) had a more positive attitude toward Washington. During this brief opening, New START was signed in 2010. But then Medvedev stepped down to make way for Putin's return.

After Putin's reelection in 2012, US–Russian relations went into free fall. There were large demonstrations in Russia against Putin after the election, and he apparently believed they were organized and financed by the United States. When the new US ambassador, Mike McFaul, arrived, it was reported in the Moscow media that he was sent by Obama to help overthrow Putin.

By this time, Putin had decided to give up on the West and "make Russia great again" by appealing to Russian nationalism, fueled by anti-US rhetoric. In 2014, Russia hosted the Winter Olympics in Sochi and put on an impressive show to announce to the world that Russia was back. (It was later shown that Russian athletes had used illegal drugs in the Games, and Russia was barred from the 2018 Winter Games in South Korea.) Soon after, Russia began military operations in Crimea and then moved troops into eastern Ukraine. As if to make it perfectly clear to Americans that Moscow could not be trusted, the Russian government interfered in the 2016 US presidential election.

The collapse of US–Russian relations from 1997 to the present is a tragic tale. What started as a promising post–Cold War courtship, with great potential to reduce nuclear dangers, has now drastically deteriorated. NATO expansion, NATO action against Serbia, and NATO missile interceptor deployments all played a key role. Together they were seen by Moscow as signs of encroachment and disrespect for Russia and its interests.

The currently poisonous US relationship with Moscow has become a key roadblock to working cooperatively to reduce nuclear arsenals and adopt less-threatening policies and postures. We must find a way to create a constructive dialogue with Moscow on nuclear security.

"Deterrence cannot protect the world from a nuclear blunder or nuclear terrorism," George Shultz, Sam Nunn, and I (Bill) wrote in a *Wall Street Journal* op-ed. "Both become more likely when there is no sustained, meaningful dialogue between Washington and Moscow."[49]

## SENATE REJECTS THE TEST BAN

After years of atmospheric nuclear tests, in 1959 radioactive deposits were found in wheat and milk in the northern United States. As scientists and the public became aware and concerned, opposition rose against nuclear testing.

A global halt to nuclear testing has been a central, bipartisan US objective since President Eisenhower first sought a comprehensive ban. After the Cuban Missile Crisis, President Kennedy and Premier Khrushchev got close in 1963 but had to settle for a ban on tests in the atmosphere, underwater, and in space, but that allowed tests underground. This reduced radioactivity from reaching the environment, but it did nothing to stop the qualitative arms race.

Thirty years later, President Bill Clinton led a global campaign to ban all nuclear tests, and in 1996 the United Nations General Assembly adopted the Comprehensive Nuclear Test Ban Treaty (CTBT). During the ratification process, however, it ran into roadblocks, and an effective testing ban is still out of reach.

In summer 1995, the 180 members of the Nuclear Non-Proliferation Treaty (NPT) met in New York to extend the treaty. The United States and the other nuclear-armed nations wanted to extend the NPT indefinitely (i.e., forever), but states without the bomb were concerned that there had not been enough progress on nuclear disarmament as required by article 6 of the treaty. As a political down payment for indefinite extension, the nuclear-armed states agreed to seek a CTBT by 1996. The treaty was signed by 71 nations, including the United States, Russia, and China. President Clinton may have thought he'd completed the job started by Eisenhower and Kennedy, but not enough members of the US Senate shared that goal, so the treaty has not been ratified by the United States.

We note here that in retrospect the forever extension of the NPT was a mistake. States without nuclear weapons were worried that if they gave up periodic votes to extend the pact, they would lose their leverage to call for greater progress on nuclear disarmament. They were right. The main promise made in 1995, to achieve the CTBT, remains incomplete, and other than New START there has been little progress on disarmament in the past twenty-five years and many setbacks. It is this lack of progress that led in part to the rise of the ban treaty movement. There will be an international conference in 2020 or 2021 to mark the NPT's fiftieth anniversary and to review the pact's effectiveness. We expect this event to devolve into a shouting match between those with the bomb and those without. Twenty-five years ago, the states with nuclear weapons had to make their case that they were living up to their disarmament commitments to try to convince the non-weapon states to extend the treaty. Now that the pact is permanent, the nuclear states essentially ignore the NPT, and we have lost what had been a significant lever to bring international attention to the issue.

President Clinton had agreed to sign the CTBT but was concerned about getting the support from his military and from the weapons labs. If either opposed the treaty, it would be very hard to bring the Senate on board; if both opposed the treaty, it would be impossible to get ratification. I (Bill) was in favor of the treaty and told the president that I would take on the job of bringing the Joint Chiefs and the weapons labs on board. In this I had the indispensable support of Gen. John Shalikashvili, the chairman. He worked assiduously with the other chiefs, who were at first reluctant, but finally worked out an agreement that they would support it if they could get a clause that required an annual evaluation by the weapons lab directors stating that they could perform their mission without testing. I was confident that the president would agree to that, so we had the military on board. Then I called a meeting of the lab chiefs to understand their concerns. The annual certification not only eased the chiefs' concerns, but it also eased the concerns of the lab directors. So, the primary discussion with them was over the definition of the test ban: Would the ban allow low-yield nuclear tests? The lab directors wanted to be able to conduct very low-yield laboratory tests that would be useful and that they believed the Russians would conduct, because they could not be detected. My assessment was that the lab directors did not consider this a make-or-break issue.

While these internal discussions were underway, considerable pressure was coming from outside the government for a true "zero-yield" treaty. That summer, over 35,000 citizens (activated by a strong coalition of nongovernmental organizations) contacted the White House calling for a truly comprehensive nuclear test ban. This media attention and public pressure was helpful in the internal discussions that I was having with the lab directors, and they decided not to make low-yield tests a condition for support. With these two key constituencies ready to support the treaty, I was confident that we would be able to get ratification in the Senate, and I so advised President Clinton.

On August 11, 1995, President Clinton announced his support for a "true zero-yield" test ban. Had we submitted the treaty for ratification quickly, I am confident that we would have succeeded. But for reasons having to do with UN actions as well as a lack of urgency on the part of the administration, the treaty languished for years before being submitted for ratification. By that time, Gen. Shalikashvili and I were out of government, the agreements that we had secured had begun to waver, and new people in the military and weapons labs were undermining the treaty with key people in Congress.

When Clinton finally sent the CTBT to the Senate, I (Tom) was working at the Union of Concerned Scientists as director of the Global Security program. CTBT ratification was our top priority. But the Senate was under Republican control and Senator Jesse Helms (R-NC), whose Foreign Relations Committee has jurisdiction over treaties, said the CTBT was a low priority and that he would only look at it after the administration sent him two unrelated pacts: the 1997 amendments to the ABM Treaty (which delineated the point at which a missile defense would be considered strategic or nonstrategic) and the Kyoto Protocol on climate change, both things Helms wanted to kill. In other words, Helms was holding the test ban treaty hostage.

Politicians had always played politics with nuclear weapons, but this was a whole new ball game. During the Cold War, once agreements were reached, the Senate, after vigorous debate, would ratify them. But this was the first Democratic nuclear treaty brought to a Republican Senate after the Cold War's end.[50] Republicans no longer felt constrained by Cold War standards of conduct, and now saw themselves free to play politics with nuclear security. Moreover, with the nuclear arms race over, the public had turned its attention to other issues.

And then the bottom fell out of the Clinton presidency. In January 1998, the story broke that Clinton had been having an affair with Monica

Lewinsky. The House of Representatives impeached Clinton in December. The Senate acquitted him in February 1999. Republicans needed 67 votes to remove Clinton from office and came up short. The legal ordeal was over, but the political damage was done.

Republicans, frustrated that Clinton had slipped through their fingers, took aim at his congressional priorities. In July 1999, all 45 Democratic senators signed a letter urging Helms to conduct hearings on the CTBT and report it to the full Senate for debate. When Helms snubbed the request, Senator Byron Dorgan (D-ND) threatened to hold up Senate business unless the treaty got a vote. "This is going to be a tough place to run if you do not decide to bring this issue to the floor of the Senate and give us the opportunity to debate [the CTBT]," he warned on September 8.[51]

What Helms knew, but the Democrats did not, was that the Republicans already had the votes lined up to kill the treaty. James Schlesinger, who once headed the Defense and Energy departments, and Senator Jon Kyl (R-AZ) had been working quietly behind the scenes to organize the opposition, and they only needed 34 votes. They had that and more. Meanwhile, the Democrats believed that if there was a vote, Republicans would not carry through with their threat to vote against the CTBT because it enjoyed broad public support. A 1998 national poll found that 73 percent of Americans favored the treaty. Surely, the Democrats thought, Republicans would not run the risk of voting down such a popular agreement.

Confident he could defeat the treaty, Helms proposed a vote on the CTBT within two weeks. Democrats unwisely accepted the deal. This was a massive intelligence failure on behalf of the White House, Democratic senators, and outside groups like mine (Tom's), which did not even see the trap we had just walked into.

With the vote just days away, President Clinton, Secretary of State Madeleine Albright, and Secretary of Defense William Cohen finally launched a high-profile effort to win Senate support for the treaty. I (Tom) went to the White House to see President Clinton make a speech in support of the treaty and was impressed with his command of the arguments. But the effort was too little, too late.

The day before the vote, it was clear that the treaty would go down. To prevent a damaging blow to US credibility, a bipartisan group of 62 senators wrote to the Republican leadership "in support of putting off final consideration until the next Congress." Senate Majority Leader Trent Lott (R-MS) and Minority Leader Tom Daschle (D-SD) were on the verge of an agreement to postpone the vote, but senators Helms, Paul Coverdell (R-GA), James Inhofe (R-OK), Kyl, and Bob Smith (R-NH) reportedly raced to Lott's office to tell him that they would block any agreement to postpone the vote.[52]

On October 13, 1999, the US Senate rejected the CTBT by a vote of 48 in favor and 51 against. (The treaty needed 67 votes to pass.) Watching the Senate vote down the CTBT was one of the most difficult things I (Tom) ever had to do.

Twenty years later, the CTBT still sits in limbo. The United States has not tested a nuclear weapon since 1992 and sees no compelling need to. But the treaty to ban testing for the rest of the world cannot come into force until the United States ratifies it. The United States led the world to negotiate the treaty, but then could not get its own house in order to approve it. This was a huge setback for US leadership and international credibility, as well as for efforts to reduce nuclear dangers.

If the nation with the most sophisticated nuclear arsenal, that has conducted the most nuclear tests (more than a thousand), and has the best simulation capabilities cannot approve the CTBT, why should anyone else? From the Russian point of view, the fact that Washington has not ratified the treaty must raise suspicions that the United States may one day resume testing, and that Moscow should be ready for such a possibility.

The CTBT is a cautionary tale of how hard it has become, because of political polarization, to get Democratic treaties through the Senate.

Unfortunately, things can get worse for the CTBT. In 2019, the Trump administration accused Russia of violating the treaty. "The United States believes that Russia probably is not adhering to its nuclear testing moratorium in a manner consistent with the 'zero-yield' standard," Robert Ashley, director of the Defense Intelligence Agency, said on May 29.[53] The administration provided no supporting evidence, and Moscow denied the accusation.

## OBAMA GETS HALF A LOAF

During the 2008 presidential election, it seemed we were heading into an alternate reality where there was bipartisan support for deep nuclear arms reductions. "The Cold War ended almost twenty years ago," said Senator John McCain (R-AZ), the GOP nominee, "and the time has come to take further measures to reduce dramatically the number of nuclear weapons in the world's arsenals."[54]

Soon after his inauguration, on April 5, 2009, President Barack Obama gave his first major foreign policy speech in Prague. He said, "So today, I state clearly and with conviction America's commitment to seek the peace and security of a world without nuclear weapons. I'm not naive. This goal will not be reached quickly—perhaps not in my lifetime. It will take patience and persistence. But now we, too, must ignore the voices who tell us that the world cannot change. We have to insist, 'Yes, we can.'"[55]

The speech created great hope in the United States and the international community that progress would be made on disarmament. Obama received the Nobel Peace Prize that October in part for the commitments he made in Prague. Those commitments included (1) preventing nuclear terrorism and promoting nuclear security; (2) strengthening the nonproliferation regime; (3) supporting the peaceful use of nuclear energy; and (4) reducing the role of nuclear weapons.

The Obama administration made real progress on many of these issues. It concluded the New START Treaty with Russia in 2010. It organized four successful Nuclear Security Summits that convened more than fifty world leaders to take tangible steps to prevent terrorists from gaining access to nuclear weapons and materials. It negotiated a comprehensive, long-term deal with Iran to prevent it from obtaining a nuclear weapon.

But the administration fell short in many areas. To give one example, Obama had promised to seek Senate ratification of the CTBT but did not mount a serious effort to do so. Obama said the United States would seek a new treaty to end the production of fissile materials for use in nuclear weapons but did not follow up. The bruising partisan battle and close vote on the ratification of New START, which should have sailed through ratification, caused Obama to back off his earlier, ambitious program on arms control. After New START, the administration had planned to seek another reduction agreement with Russia but was thwarted by Moscow due in part to its concerns about US missile defense plans. Obama chose to expand the anti-missile system in Alaska and California that had been fielded by President Bush and to field his own new system on Russia's doorstep in eastern Europe.

In addition to these setbacks, the Obama administration started an excessive program to rebuild the US nuclear arsenal. Instead of simply rebuilding what we need to maintain a strong deterrent against Russia, the Pentagon took over the project and developed a plan to rebuild all parts of the arsenal as if the Cold War had never ended. Yes, Obama had promised in Prague that "as long as these weapons exist, the United States will maintain a safe, secure and effective arsenal to deter any adversary." And yes, in exchange for Republican votes on New START, Obama had promised to rebuild and maintain the arsenal. But his administration's $2 trillion nuclear plan (over thirty years) took it all to the extreme.

Moreover, to get New START passed, Obama also agreed not to limit missile defenses. The United States issued a nonbinding unilateral statement explaining that US missile defenses "are not intended to affect the strategic balance with Russia" but that the United States intends "to continue improving and deploying its missile defense systems in order to defend itself against limited attack." The Senate's instrument of ratification repeated its intention "to deploy as soon as is technologically possible an effective National Missile Defense system capable of defending the territory of the United States against limited ballistic missile attack (whether accidental, unauthorized, or deliberate)."[56]

We strongly believe that if the $2 trillion nuclear modernization program is the ultimate price for New START, the treaty will not have been worth it. New START, although valuable, made only modest cuts to the nuclear arsenals. By approving the excessive rebuild plans and reinforcing bipartisan support for missile defense, we are locking ourselves into brand-new and dangerous weapons for the next fifty years. "They agreed to spend whatever it took to keep the ICBMs and the B-52s ready to fly for another full generation," wrote MSNBC host Rachel Maddow in her book, *Drift: The Unmooring of*

*American Military Power.* "Settle in, Missileers, it's gonna be at least another few decades."[57]

How did Obama's call for a world without nuclear weapons turn into the rebirth of the bomb? Let's start with what Obama did right. He came into office with a clear vision of where he wanted to go on nuclear weapons and had a plan to get there. He used his first foreign policy speech to lay out his nuclear agenda and show the world that the bomb was a high priority. He set an ambitious agenda, and he got off to a productive start.

But then Republicans in the Senate made New START a partisan political fight rather than provide the no-brainer bipartisan support it deserved. The treaty barely passed 71-26 (it needed at least 67 votes, or a two-thirds majority required for all treaties). Had this been a Republican treaty, it would have easily sailed through. As it was, ratification was difficult, and the Obama administration had to pay a high price on nuclear modernization. Moreover, after New START, the administration had no bandwidth left to take on Senate approval of the CTBT. As Brent Scowcroft, national security advisor under presidents Ford and George H. W. Bush, told ABC News, "I just don't understand the opposition . . . to play politics with what is in the fundamental national interest is pretty scary stuff."[58]

And after New START, Russia balked at deeper reductions, in part because of Obama's support for missile defense both on the US West Coast and in Europe. Without a dance partner in Moscow, the administration would have had to make one-sided reductions, which Obama was not willing to do, particularly after Russia's 2014 annexation of Crimea.

The New START experience points to a larger problem: Democratic presidents tend to support arms control, but it has become next to impossible for a Democrat's arms control treaty to get at least 67 votes in the US Senate. Conversely, Republican presidents, who have a long history of supporting arms control, have now turned against it. As a result, unless there is a change in this partisan behavior on existential nuclear issues, there may be no way to get future arms control treaties through the Senate.

Some of Obama's problems were self-inflicted. While the president was looking elsewhere, it was his own bureaucracy that failed to keep the nuclear rebuild in check and let it balloon into a $2 trillion behemoth. It was his own team that pushed missile defense expansion. People are policy, and even good policy can be foiled by the wrong staff. Left to itself, the Pentagon—and the Air Force, in particular—chose, in effect, to build a modern version of the Cold War nuclear arsenal rather than think through what modernization was really needed. The biggest roadblock to enacting Obama's nuclear agenda turned out to be the president's own team and his own moderation.

In the summer of 2010, after the first Nuclear Security Summit, New START, the Prague speech, and the Nuclear Posture Review (NPR), Obama's chief of staff Rahm Emanuel pulled the president aside and told him he was spending too much time on nuclear policy. "You said you wanted to run for president to pass health care. And we're in the fight of our lives," Emanuel said to the president, as told to Jon Wolfsthal, "and you can't spend as much time on nukes if you want to do this." "And the president heard that," Wolfsthal said. "So, we spent much less time after 2010. And we passed health care."[59]

When Wolfsthal left government in April 2012, there were others still working on nuclear policy, but there was no single person whose main job was focused on achieving the nuclear agenda in the Prague speech. Meanwhile, the Iran nuclear deal negotiations took up significant time and were not completed until 2015.

While Obama focused elsewhere, the nuclear agenda drifted. "At a certain point we lost control of this," Obama's deputy national security advisor Ben Rhodes said. "We did not intend to set this snowball rolling down the mountain in 2009. We intended to do something that seemed to us to be a responsible way . . . to have a safe infrastructure around our nuclear weapons and get New START ratification done. But what ended up happening . . . is it did become this giant snowball. And it is insane and unnecessary and wasteful and an offensive use of prioritization that we would be spending this amount of money to modernize nuclear weapons when we have threats from climate change and much more pressing threats."[60]

Furthermore, when Obama tried to change nuclear policy, he encountered at every turn a built-in body of opposition in the uniformed military who supported nuclear weapons. "I think this is a fair criticism of the president and he should own it," said Wolfsthal. "On almost every occasion, when it came to the Prague agenda on reductions of nuclear weapons, on changing nuclear doctrine, he chose not to take the most ambitious, most radical step because he wanted to preserve his leverage and his relationships and his capital with other parts of his administration for other issues. Maybe it was Iran, drawdowns in Afghanistan, maybe it was other pieces, but he had things teed up and chose actively not to take those steps. And that I think is a recurring theme for the president; that these things were available to him and, time and again, his moderation led to maintaining the status quo."[61]

Obama, like Truman, started out with good intentions and wound up in a very different place. Both presidents took their eye off the ball when it mattered most. If left to itself, the nuclear bureaucracy—Pentagon planners, defense contractors, congressional champions, and think tank boosters—will keep the contracts and the money moving. That is the powerful current of conventional thinking, and any president who wants to make a difference has to swim against it. "Changing the status quo takes work. There's not a built-in constituency that's going to applaud you for it, and you have entrenched interests that believe you're wrong," said Wolfsthal.[62]

Leah Greenberg, co-founder of the mass movement organization Indivisible and one of *TIME* magazine's 100 most influential people for 2019, told Ploughshares Fund president Joe Cirincione during a 2019 interview that one of the hard lessons learned from the Obama years was that, after the election, many of Obama's supporters thought the job was done. They went home. But those opposed to Obama's ideas, including on nuclear weapons, mobilized. Today, Greenberg said, we have to think about how to build a coalition that will still be around to push for reforms once a new president is in office. "Because what Democrats really were missing in 2009 was an amped-up grassroots that was really having their back as they were debating these big policies that they were trying to move."[63]

---

Why do we still have the bomb? Because, after seventy-five years, the bomb is the deeply embedded default setting. Changing the setting takes leadership from the very top, and it takes focused, sustained attention over many years. Given the lack of public pressure demanding attention to nuclear issues and the vast array of other prominent issues any president must deal with, it is no wonder the bomb is alive and well. The bomb is a survivor, and it thrives in the shadows.

Uprooting the bomb, and related missile defenses, will take sustained effort by a committed president. Unless the public demands this from their president, it is unlikely to happen. If we are ever lucky enough to get another president who comes into office with the right nuclear agenda, we must remember that without strong public support the president's attention will be diverted. It is not enough for a president to want to do the right thing. The public must still demand it.

As President Abraham Lincoln once said, "In this age, in this country, public sentiment is everything. With it, nothing can fail; against it, nothing can succeed."[64]

# CHAPTER 10

# THE ATOMIC
# *TITANIC*

*The story of nuclear weapons will have an ending, and
it is up to us what that ending will be. Will it be the end
of nuclear weapons, or will it be the end of us?*
—Beatrice Fihn, accepting the 2017 Nobel Peace Prize[1]

The RMS *Titanic* was steaming through the North Atlantic from Southampton to New York on April 15, 1912, when the impossible happened. The unsinkable ship struck an iceberg and sank, killing more than 1,500 people. It was one of history's deadliest peacetime marine disasters.

Until the moment of impact, all was fine. In fact, the false impression that the ship was "unsinkable" may have caused overconfidence among the crew that led them to take greater risks than they otherwise would have.

Similarly, the fact that we have lived with nuclear weapons for seventy-five years has lulled us into a false complacency. To those who say, "We have had nukes this long and all is well, so what's the problem?" we say, "The extreme dangers are there, lurking just beneath the surface, and you don't even see them. Look closer and you will."

"So, we're almost like passengers on the *Titanic*," former California governor Jerry Brown said in 2019. "Not seeing the iceberg up ahead but enjoying the elegant dining and the music. The business of everyday politics blinds people to the risk. We're playing Russian roulette with humanity. And the danger and the probability [are] mounting that there will be some nuclear incident that will kill millions, if not initiating exchanges that will kill billions."[2]

We agree with Governor Brown. US nuclear policy is a disaster waiting to happen, and it could be only a matter of time until our luck runs out. The Cold War is over, and we have had thirty years to rethink US nuclear policy. Yet through the administrations of George H. W. Bush, Bill Clinton, George W. Bush, Barack Obama, and now Donald Trump, we have failed to learn the correct lessons from the Cold War. Each president sought to rearrange the deck chairs on the *Titanic*, but none set a new course away from the icebergs.

---

Unlike all other instruments of war, nuclear weapons are the president's weapons. Only the commander in chief can authorize their use. With this authority comes the responsibility for presidents to take the lead on transforming nuclear policy so these weapons are never used again.

And unlike all other nations, the United States bears the greatest global responsibility to lead the world away from the bomb. America brought the bomb into the world and has in the past been the global leader in efforts to control it. The United States must once again take up the essential cause of nuclear disarmament.

Whoever is elected in 2020, the next president of the United States will have a unique opportunity to reshape US nuclear policy. The seventy-fifth anniversary of the bomb and the fiftieth anniversary of the Nuclear Non-Proliferation Treaty (NPT) will shine a bright light on our current policies that are broken and badly in need of repair.

As we have seen in these pages, the defining error of US nuclear policy is that it is focused on the wrong threat. We are preparing for a first strike from Russia that is *very* unlikely; what is not so unlikely is that we will *blunder* into a nuclear war. Yet by preparing for the surprise first strike, we actually make the blunder more likely. Our misguided policies like sole authority, first use, and launch on warning are extremely dangerous, particularly when combined with the old dangers of false alarms, the new dangers of cyber threats, and the ever-present dangers of an unstable president—even if that instability is temporary as a result of medication or overdrinking.

Meanwhile, instead of fixing these glaring problems, the United States is doubling down on them by spending more than a trillion dollars to rebuild its nuclear arsenal as if the Cold War had never ended. We could increase our security while at the same time saving hundreds of billions of dollars by shifting to a policy of second-strike retaliation and phasing out the weapons that are most prone to be used first and quickly, such as ICBMs.

In addition, the United States and Russia are rushing into a nuclear arms race that neither side can win. Washington and Moscow are tearing down the arms control structures that served them so well over the last fifty years. It is vital to our security that these structures be preserved, and that we continue the process of reducing nuclear arsenals. But we cannot get there unless we address missile defense, the third rail of nuclear arms control.

As the Cold War neared its end, Reagan and Gorbachev came to realize that the vast nuclear arsenals we had were not necessary for our security and in fact posed existential dangers to both countries. They began to reverse the nuclear buildup, even discussing the total elimination of nuclear weapons. Perhaps the greatest legacy of those two leaders was their oft-stated judgment: "A nuclear war cannot be won and must never be fought."

That assertion is an excellent starting point for crafting a new nuclear policy: one that recognizes that neither Moscow nor Washington will initiate an unprovoked nuclear attack on the other; that greatly lowers the danger of nuclear war by a technical or political miscalculation; that significantly reduces the number of nuclear weapons; and in which no nation uses nuclear weapons to threaten another nation. All nuclear-armed states should move quickly to ensure their security without nuclear weapons and, in time, eliminate their nuclear stockpiles.

The United States needs to learn the right lessons from the Cold War—the real danger is that we could accidentally bring on a nuclear catastrophe. We need to change gears and design a nuclear force and policies to minimize this danger. This will save money, prevent a new arms race, and make us all safer.

Significant changes in US nuclear policy will not happen without broad public awareness and support. The issue of presidential sole authority—who has his or her finger on the button—has become the most resonant nuclear policy issue in forty years, since the nuclear freeze movement in the 1980s.

We are all on the atomic *Titanic*, and the ship is headed for a hidden iceberg. To steer away from disaster, the United States must make major changes to its nuclear policies.

## TOP TEN RECOMMENDATIONS

Here are our top ten recommendations for a safer world:

**1.  End presidential sole nuclear authority. Retire the "football."**
The US Constitution gives Congress the sole authority to declare war. Certainly, using nuclear weapons to attack another country would be the ultimate expression of waging war, so that authority lies with Congress. During the Cold War, policies evolved that effectively set aside the Constitution by giving to the president the sole authority to launch nuclear weapons. We no longer live in a world—if we ever did—where one person should have the absolute power to end life on earth.

Presidential sole authority was first adopted to place the decision to use the bomb firmly in the hands of civilians, which we fully support. But there is no need to limit this authority to *just one* civilian. In the case of first use, we support current legislation to require a declaration of war by Congress that specifically authorizes a nuclear attack before the president can use nuclear weapons. First use should require the *shared* authority of the legislative and executive branches.

In the case of retaliation to a nuclear attack, sole authority may be justified, but only after such an attack has been confirmed by actual detonations (in other words, launch on warning would be prohibited; see below). At this point, the nation would be in a state of war, and the president would have the authority to launch nuclear weapons without seeking the approval of Congress. Even in that case, we believe that the president should try to consult with senior advisors before launching a nuclear retaliation.

Thus, sole presidential authority should be allowed *only* in retaliation to a confirmed nuclear attack on the United States (or an ally covered by our extended deterrent). As such, there would be no need for the president to launch nuclear weapons quickly, within minutes. There would be time for a measured response. If a nuclear attack appears to be underway against the United States, the president, rather than worrying about launch options, should use these precious minutes to get to a secure location to establish communications with civilian and military advisors. There would no longer be a need for a military aide to follow the president, 24-7, with the emergency satchel. It is time to retire the football.

## 2.  Prohibit launch on warning.

As discussed, launching nuclear weapons on *warning of attack* but before the attack can be unambiguously confirmed is just too dangerous to be contemplated. There is simply no realistic scenario that justifies a decision to launch nuclear weapons within minutes given the inherent dangers of doing so. Given the tremendous consequences of the decision (the fate of the world) and the mind-crunching time pressure (ten minutes or less) to make such a decision, it is not worth the risk. In addition, the president would likely be working with incomplete information (Kennedy, Cuban Missile Crisis); could be under the influence of alcohol (Nixon); could be responding to a false alarm caused by equipment malfunction (Carter); or could have a predisposition to act impulsively on what could be a false alarm caused by a cyberattack (Trump).

Simply put, the stakes are far too high for the United States to ever launch nuclear weapons unless we are absolutely sure that we are under nuclear attack. Once an attack is confirmed, the United States would still have a survivable second-strike force based on submarines at sea. We should never rush into nuclear war, even in retaliation.

### 3. Prohibit first use.

Given US conventional superiority, we believe that no rational president would use nuclear weapons first, in any scenario, and thus US threats to do so are not credible. Against a nuclear-armed state like Russia, first use would be suicide in the face of assured retaliation. Against a nonnuclear state, first use would start a race among such states to go nuclear, make the United States into an international outcast, and go against fifty years of US nonproliferation policy. How can we possibly convince other states that they do not need nuclear weapons if the United States itself says it needs them for nonnuclear threats? We would be stepping back in time to a nuclear wild west.

Nuclear weapons serve no practical military purpose other than to deter their use by others. The United States can deter and respond to other threats (biological, chemical, and conventional) with its unmatched conventional arsenal. In the past, some US allies have urged the US to retain a first-use policy. US allies need to be reassured that a policy of no first use does not undermine Washington's commitment to their security.

The United States can make a policy of no first use more credible by adjusting its nuclear forces accordingly. ICBMs have no role other than as first-strike weapons (they would not survive a first strike by Russia), and should be phased out. Until then, ICBMs should be taken off alert in a verifiable way, such as storing warheads separately from missiles. The US strategic bomber fleet is already de-alerted, with bombs stored in bunkers. Washington and Moscow could work together to develop ways to reduce the first-strike threat from submarines at sea by, for example, limiting their deployment areas away from each other's coasts.

Finally, a credible US no-first-use policy could encourage Russia to follow suit and back away from its hair-trigger launch status. The danger of Moscow blundering into nuclear war because of its first-use and high-alert policies is probably even greater than in the United States. Thus, it is in the US national security interest to do what it can to nudge Russia away from first use and into a second-strike-only posture.

We support current legislation that would make it US policy to not use nuclear weapons first. This would supplement legislation to limit presidential sole authority—a "belt and suspenders" approach to prohibiting first use. A no-first-use policy could also be achieved by executive order.

### 4.   Retire all ICBMs and scale back the nuclear rebuild.

Retiring ICBMs would solve a number of problems for US nuclear policy. It would reduce the pressure to launch on warning and "use them or lose them"; make US no-first-use policy more credible; and save hundreds of billions of dollars that could be redirected to higher-priority projects.

Without first use, ICBMs would have no legitimate purpose. Most of them would be destroyed by a Russian first strike and would be unnecessary in any other scenario. The ICBMs are simply not needed for an effective response, which would be carried out by submarine-based weapons. ICBMs are not worth keeping as a "nuclear sponge," as this only increases the nuclear danger to the US Upper Midwest and surrounding states. Drawing a nuclear attack *to* the United States rather than *away* from it makes no sense.

US plans to spend about $150 billion to build a new generation of ICBMs are not only a waste of taxpayer money, but deploying those weapons would make us less safe. The ICBMs are, at best, extra insurance that we do not need; at worst, they are a nuclear catastrophe waiting to happen.

The ICBMs are part of a larger effort to spend about $2 trillion on rebuilding and maintaining the US nuclear arsenal over the next thirty years. This effort is excessive. The United States should build only the weapons it needs for second-strike deterrence and should not go beyond that for obvious reasons: the weapons are expensive and dangerous.

The US nuclear-armed submarine force alone is sufficient for assured deterrence and will be so for the foreseeable future. But as technology advances, we have to recognize the possibility of new threats to submarines, especially cyberattacks and detection by swarms of drones. The new submarine program should put a special emphasis on improvements to deal with these potential threats, ensuring the survivability of the force for decades to come.

In the context of no ICBMs, we believe that a fleet of ten new nuclear-armed submarines would be more than adequate to meet our country's deterrent needs. The firepower on board just five or six survivable submarines would be enough to destroy the vital elements of state control, power, and wealth in Russia, China, and North Korea. In fact, just *one* boat can carry enough nuclear weapons to place two thermonuclear warheads on each of Russia's fifty largest cities.

The Trump administration is fielding new "low-yield" nuclear warheads on Trident submarine-based missiles. These dangerous weapons are a bad solution to a nonexistent problem. The United States can deter the unlikely Russian use of its low-yield bombs with its current arsenal. There are no "gaps" in the US deterrent force, and there can be no doubt in Russia's mind that the United States is serious about maintaining an unambiguously strong nuclear deterrent.

Trump's program calls for development of a new bomber, the B-21, with improved stealth capability. We support that program because it is a useful addition to our conventional forces and because it provides backup should the submarines ever suffer a temporary problem that raises a question about their capability. This is not likely, but the bomber force is an insurance policy for that contingency.

Moving from first use to second-use assured retaliation means that the command and control system can shift from quick launch options to providing more decision time for the president. We do *not* need weapons on high alert and the ability to launch on warning of attack. But we do need a survivable system that protects the president and his or her ability to issue orders under the most stressful conditions imaginable.

The United States should prioritize command and control modernization over rebuilding its nuclear weapons. The president should not feel rushed into a launch decision, and we should seek to extend the time frame well beyond an attack. This would better allow the president to reassess the post-attack situation and prudently direct the operations of surviving forces.

As the United States plans for the future of the nuclear arsenal, we can move to a smaller but more secure second-strike force whose sole purpose is to deter nuclear attack. We do not need to spend hundreds of billions more in a dangerous and futile attempt to "prevail" in a nuclear conflict.

By phasing out ICBMs and building fewer new submarines and bombers, we estimate that the United States could save *at least* $300 billion ($10 billion per year) and still field a formidable deterrent force.[3]

### 5.   Save New START and go farther.

The New START Treaty, negotiated by the Obama administration and signed by the United States and Russia in 2010, is the last major agreement still in force limiting nuclear arms, and it expires in February 2021. It can be renewed for five years, but only if Washington and Moscow agree. President Trump may do nothing or oppose extension. In that case, a new president who takes office in January 2021 would have just weeks to pick up the pieces.

President Trump should resist efforts to complicate New START extension. Yes, Russia is developing new nuclear weapons not covered by the treaty, and yes, China is not in the treaty at all. Options to address these issues can be explored for a follow-on agreement but will simply take too long to consider by early 2021. In fact, we believe that the proposal to include these issues in the New START extension is simply a spoiling effort by those opposed to New START.

Abandoning New START would be a tragic error that would throw gasoline on the arms race fire. New START has served the United States well, and there are no indications of Russian violations. It deserves to be renewed. It limits the number of nuclear weapons that Russia can aim at the United States, and it gives Washington confidence that Moscow will not expand its arsenal quickly. Without New START, those assurances disappear. New START is also the vehicle for the last remaining dialogue we have with Russia on nuclear weapons. Indeed, one of the most important reasons for strategic arms treaties is to maintain a continuing dialogue on these issues that affect the survivability of both of our countries.

However, it would be a mistake for Congress to allow the Trump administration to hold the treaty hostage and, as ransom, demand support for the administration's excessive nuclear modernization program. As much as we want New START extended, this is a price not worth paying.

Even if New START is extended, it cannot be the end of the arms control road. Both Moscow and Washington would still retain enough nuclear firepower to destroy the world for themselves and everyone else. We must continue to drive numbers down so that if nuclear war does break out, it does not result in the end of civilization. Climate science tells us that a hundred or more high-yield nuclear warheads detonated on large cities could so damage the climate as to cause a major deterioration of life on earth. We cannot consider the arms reduction process truly successful until it brings the numbers down to this level and below.

After New START, President Obama had planned to negotiate a follow-on treaty with Russia to reduce deployed strategic nuclear forces by one-third, or down to about a thousand warheads. This could serve as the starting point of renewed talks with Moscow.

As part of this effort, presidents Trump and Vladimir Putin should announce a joint declaration reaffirming that a nuclear war cannot be won and must never be fought. This would renew the 1985 Reagan-Gorbachev statement that Americans and Russians received positively as the beginning of an effort to reduce risk and improve mutual security. A joint statement today would clearly communicate that despite current tensions, leaders of the two countries possessing more than 90 percent of the world's nuclear weapons recognize their responsibility to work together to prevent a global catastrophe. This could also lead other nuclear states to take further steps to reduce nuclear risk. The timing of such a statement would also signal Washington and Moscow's commitment to build on past progress toward disarmament.

We fully understand that US–Russian relations are at a historic low and that building political support for new negotiations will be challenging. But the importance of reducing global nuclear dangers should be clear to all and transcend political calculations on nonnuclear issues. Whatever their other areas of disagreement, leaders in Washington and Moscow should always be willing to talk about avoiding nuclear war.

### 6. Limit strategic missile defenses.

It will be next to impossible to continue with significant arsenal reductions without real limits on strategic missile defenses. As long as Russia sees US missile defenses as a threat to its ability to retaliate after a US first strike, Moscow will refuse to negotiate major new arms reductions without limits on US defenses. And as long as the US Congress remains under the mistaken belief that these systems can be made to work effectively, it will refuse to limit national defenses.

After spending well over $300 billion on strategic missile defense programs since 1983, the more Washington spends to deploy missile interceptors and develop space weapons, the more Moscow will resist reducing its nuclear forces and the more Beijing will build up. This will put an inevitable brake on how far we can get in the reduction of nuclear weapons. Russia and China are also concerned about US regional missile defenses, such as US sea- and land-based interceptors in Europe, Japan, and South Korea.

One way to address this problem is to seek to convince the US Congress that long-range missile defenses are not as effective as represented. A main reason why Congress is under the delusion that US interceptors are effective is that they have not been tested against realistic threats. The George W. Bush administration deployed a deeply flawed system, and it has regrettably won bipartisan support. But we should not expand it or add to it in any way until those systems have been run through independent, realistic tests, including real-world countermeasures. We do not believe that the US strategic missile defense system will pass such realistic tests. We should not continue to spend billions on systems that are not effective and that lead us to a false sense of security.

We are making strong statements about missile defense, but the future of the US program need not be based on our negative assessments. We also need independent technological assessments. Congress should fund an independent organization to review the numerous proposed anti-missile technologies. A 1987 report by the American Physical Society (APS) on the feasibility of directed energy weapons helped scale back the SDI program. A similar study by APS on current anti-missile systems would be a boost to congressional oversight and public understanding of these complex technologies.

Ultimately, Washington and Moscow will need binding limits on long-range missile interceptors, as existed in the ABM Treaty until it was abandoned by the George W. Bush administration in 2002. It is hard to imagine securing Moscow's support for further arsenal reductions without such binding limits.

## 7.  Don't wait for treaties.

As much as we support the use of legally binding treaties to secure limitations on nuclear weapons and missile interceptors, unfortunately, we can no longer count on this approach to succeed. The Comprehensive Test Ban Treaty (CTBT) was rejected in 1999 by a Republican Senate even though it would clearly serve US interests. The New START Treaty barely made it through a Democratic Senate and did so only after the Obama administration promised it would rebuild nuclear forces and not allow limitations on missile defenses. Meanwhile, Republican presidents, who have a long history of supporting arms control, have now turned against it. As a result, there may be no way to get future arms control treaties through the Senate.

The sad reality is, for the foreseeable future, we cannot count on the US Senate to produce 67 votes for arms control measures, even if those measures are clearly serving US national security interests.

But these issues are too important to sit by and wait for a change in the composition of the Senate. If political support in the Senate does not exist for prudent agreements on arms reductions, the next president should pursue them through executive agreements and other means that do not require ratification by the Senate. This is far from ideal but still better than nothing.

It is true that political agreements may be more vulnerable to being reversed by the next president, as shown by Trump's withdrawal from Obama's nuclear deal with Iran. But it is also true that presidents can just as easily withdraw from treaties, as Trump did with the INF Treaty and Bush did with the ABM Treaty. Treaties are not our only path to progress, and if partisan politics make treaties impossible, then we must pursue other means.

## 8.  Engage diplomatically with North Korea and Iran.

The next president should seek to use diplomacy to contain nuclear programs in North Korea and Iran. Such efforts are essential to restraining regional arms races as well as for allowing truly deep reductions in US and Russian forces. There are no viable military solutions to either situation, and such actions should be avoided.

On Iran, the next US president should reenter the Iran nuclear agreement negotiated by Tehran and the United States, Russia, China, and the

European Union. According to international inspectors, Tehran was complying with the 2015 Joint Comprehensive Plan of Action (JCPOA) when the Trump administration withdrew from the agreement in 2018. Trump's ill-advised action has only served to set back the goal of preventing Iran from developing nuclear weapons. The next president should seek to rejoin the JCPOA and begin to rebuild good relations with Tehran toward the possibility of negotiating a follow-on agreement to extend international limits into the future.

North Korea, by contrast, already has nuclear weapons and poses an even tougher diplomatic challenge. We support President Trump's effort to engage directly with North Korean leader Kim Jong-un, but so far the administration has not demonstrated that it has a viable negotiating strategy. Rather than asking the North to surrender its nuclear arsenal at the start of a diplomatic process, the United States should seek to build a fundamentally new relationship with Pyongyang such that North Korea no longer fears unprovoked military action by the United States, and South Korea no longer fears unprovoked military action by North Korea. Such a transformed relationship will take time and patience.

The Nuclear Non-Proliferation Treaty (NPT) will mark its fiftieth anniversary in summer 2020. The treaty and its 190 member-states play an important role in stopping the spread of the bomb to additional countries. The great majority of the states without nuclear weapons support the UN treaty to eliminate the bomb and are frustrated that the United States and Russia have not made more progress on nuclear disarmament and are, in fact, abandoning arms reduction treaties and rebuilding their nuclear forces. In this regard, Moscow and Washington could build much-needed support for the NPT process by extending New START and working to bring the CTBT into force.

**9. Bring the bomb into the new mass movement.**
As we saw in chapter 9, even if a new president is committed to transforming US nuclear policy, once in office, she or he will face tremendous political headwinds and institutional resistance to change. To overcome these barriers, a new president will need to come into office with a clear plan of action that

can be implemented quickly, and put the right staff in the most influential positions to enact that policy.

As in the case of President Obama, once the election is over, there will still be a need for pressure from outside organizations and American voters. There should be a powerful outside constituency to remind the president of promises made and that there will be a political cost if progress is not achieved.

Public education is essential (and is the reason we wrote this book). But we know this is just the beginning. There should also be tweets, op-eds, articles, podcasts, speeches, conferences, YouTube videos, TV shows, documentaries, and mainstream movies. The goal is to change the way the public and popular culture view nuclear weapons. The weapons need to be seen not as an asset, but as a liability—indeed, an existential danger.

The Nuclear Freeze campaign peaked thirty-eight years ago with a million people protesting the arms race in 1982 in New York City's Central Park. We do not have a similar movement today, but we do have a transformed mass movement that was born soon after the election of President Trump. This popular uprising is focused on women's rights, immigration, justice, democracy, war prevention, gun control, and the environment; it could also focus on nuclear disarmament.

We need to bring the bomb into the new mass movement. Highly effective organizations like Indivisible, MoveOn, and others are leading the way. We stand ready to help however we can.

**10.  Elect a committed president.**

Finally, it all comes down to electing a president who cares about avoiding nuclear war and changing US nuclear policy. This policy must change from the top, and only the president can drive that process. Nothing will change unless the president wants it to change. American voters need to educate themselves on what the presidential candidates think about nuclear weapons and what the candidates will do if elected. So, we ask all of you who are eligible to vote in the 2020 US presidential election and thereafter: please vote as if your life depends on it.

Today, with the Cold War thirty years behind us, the United States still has massive nuclear forces deployed to respond immediately to a surprise attack, and the president has unchecked authority to start a nuclear war. US nuclear policy is stuck in a time warp, essentially unchanged since the 1950s.

The framers of the Constitution could not have foreseen the challenges that would be posed by the atomic age, but they did understand the dangers of an unchecked president. The bomb and how we manage it have caused great damage to American democracy and security by undermining congressional authority and bestowing superhuman powers to the commander in chief.

There is no justifiable need to give any president the unilateral power to end the world within minutes. As the bomb turns seventy-five, there is no better way to reduce the danger of nuclear catastrophe than by getting rid of the nuclear button.

We wrote this book with the hope that by reframing how we think about the bomb's past, we could help change the bomb's future. But that ultimately depends on you, the reader. The public must get more actively involved if we expect our national leaders to push for a new nuclear policy. We know this is possible, because we have seen it happen before. The United States and Russia ended the first nuclear arms race because of public pressure. The UN approved a global ban on nuclear weapons because of public pressure. And we will end sole authority, stop a new arms race, and build a safer world in the same way.

Today, nuclear weapons pose *the* underappreciated danger to the very existence of our civilization. No one of us alone can alter that reality, but together we can act to give this vital issue the time and attention it rightly deserves.

# ACKNOWLEDGMENTS

This book was inspired by Bill Perry, who first opened my eyes to the fact that US nuclear policy is focused on the wrong threat. I pitched him the crazy idea of writing a book together and he agreed (will wonders never cease!). We established a great working relationship, and it was a delight and honor to write this book with him. I hope the book lives up to his high standards, and I will always be grateful for his willingness to jump into this project with me. His daughter Robin Perry was the essential binding energy that held the project together from start to finish. The whole Perry clan is a force of nature, and I am honored to know them.

I could not have co-written this book without the institutional support of Ploughshares Fund and the personal support of Joe Cirincione, president of Ploughshares, and Philip Yun, former executive director, who were champions throughout the process. I am eternally grateful.

Thanks to Rick Pascocello, formerly with Glass Literary, for believing in the book and helping us craft our proposal. Thanks to the whole BenBella Books team, especially Glenn Yeffeth, Alexa Stevenson, and Stephanie Gorton, for a great partnership.

Thanks to Bruce Blair, Samantha Neakrase, Page Stoutland, Akshai Vikram, and Zack Brown for reading draft chapters and providing essential feedback. They of course bear no responsibility for any errors herein. I was also fortunate to get top-notch research support from Alex Spire, Munnu Kallany, and Annika Erickson-Pearson. No task was too small or too big.

Finally, I thank my family for putting up with my moods and general state of distraction as the book emerged. Thank you, Sara, Jared, Natalie, Lu, and Loki; you are my light at the end of the tunnel, and I love you tons.

*—Tom Z. Collina, Washington, DC, January 2020*


When I finished my memoir, *My Journey at the Nuclear Brink*, I thought that my writing days were over. But Tom Collina convinced me that another book was necessary to explain in detail what could be done to *lower* the nuclear dangers we face. He offered to co-author such a book with me, and that was an offer I could not refuse. Tom was as good as his word, and better. He prepared the outline for the book and wrote key sections of it. We approached the problem from different backgrounds and had varying perspectives, but as we labored through the major issues, we were always able to come to an agreement about how to express our common views. Tom was the driving force behind this book, and it is clear that it could not have been written without his talent, extensive knowledge, and boundless energy.

I would also like to acknowledge the influence that Governor Jerry Brown has had on this book. He carefully read my memoir, wrote an insightful review of it for the *New York Review of Books*, and has continued to work side by side with me to advance the ideas in the book. While doing so he has confronted me many times with the fact that we are not doing enough to reduce the nuclear dangers that the book so clearly describes. While he continues to prompt our leaders to take meaningful political action, he wants a clear, irrefutable articulation of what actions are essential. This book does that.

Even with these two powerful imperatives to act, I still could not have co-authored this book without the competent and loving assistance of my daughter Robin Perry, who has been with me (literally) every step of the way. Thank you, Robin.

*—William J. Perry, Palo Alto, CA, January 2020*

# NOTES

## PREFACE

1   This scenario is loosely based on an episode of *Madam Secretary*, "Night Watch," directed by Rob Green-lea, written by Barbara Hall, CBS, May 20, 2018.

## CHAPTER 1

1   Jeff Daniels, "A group of scientists is trying to limit Trump's nuclear authority," CNBC, January 24, 2018, https://www.cnbc.com/2018/01/24/scientists-seek-to-limit-trumps-power-in-ordering-a-nuclear-strike.html.

2   Mallory Shellbourne, "Trump: Receiving nuclear codes a 'very sobering moment,'" *The Hill*, January 25, 2017, https://thehill.com/homenews/administration/316225-trump-receiving-nuclear-codes-a-very-sobering-moment.

3   Nick Allen, David Lawler, and Ruth Sherlock, "Donald Trump 'asked why US couldn't use nuclear weapons if he becomes president,'" *The Telegraph*, August 3, 2016, https://www.telegraph.co.uk/news/2016/08/03/donald-trump-asked-why-us-cant-use-nuclear-weapons-if-he-becomes.

4   Julian Borger, "Ex-intelligence chief: Trump's access to nuclear codes is 'pretty damn scary,'" *The Guardian*, August 23, 2017, https://www.theguardian.com/us-news/2017/aug/23/ex-intelligence-chief-trumps-access-to-nuclear-codes-is-pretty-damn-scary.

5   Scott Horsley, "NPR/Ipsos Poll: Half of Americans Don't Trust Trump on North Korea," *Morning Edition*, NPR, September 18, 2017, https://www.npr.org/2017/09/18/551095795/npr-ipsos-poll-most-americans-dont-trust-trump-on-north-korea.

6   Michael Beschloss, *Presidents of War* (New York: Crown Publishing Group, 2018), viii.

7   Interview with Ben Rhodes, June 17, 2019, "Press the Button: Episode 10," podcast audio, *Press the Button*, Ploughshares Fund, edited for clarity with permission, https://soundcloud.com/user-954653529/ben-rhodes-former-deputy-national-security-advisor-joins-joe-cirincione-in-conversation.

8   Interview with the authors, June 19, 2019.

9   Garrett M. Graff, "The Madman and the Bomb," *Politico*, August 11, 2017, https://www.politico.com/magazine/story/2017/08/11/donald-trump-nuclear-weapons-richard-nixon-215478.

# NOTES

10   Lawrence K. Altman and Todd Purdum, "In J.F.K. File, Hidden Illness, Pain and Pills," *New York Times*, November 17, 2002, https://www.nytimes.com/2002/11/17/us/in-jfk-file-hidden-illness -pain-and-pills.html.

11   Ed Pilkington, "Ronald Reagan had Alzheimer's while president, says son," *The Guardian,* January 17, 2011, https://www.theguardian.com/world/2011/jan/17/ronald-reagan-alzheimers-president-son.

12   Andrew Cockburn, "How to Start a Nuclear War," *Harper's Magazine*, August 2018, https://harpers.org/archive/2018/08/how-to-start-a-nuclear-war/.

13   Jim Acosta and Kevin Liptak, "Sources to CNN: During Puerto Rico visit, Trump talked about using nuclear football on North Korea," CNN, March 28, 2019, https://www.cnn.com/2019/03/28/ politics/donald-trump-nuclear-football-puerto-rico-north-korea/index.html.

14   Philip Rucker, "Trump to North Korean leader Kim: My 'Nuclear Button' is 'much bigger & more powerful,'" *Washington Post*, January 2, 2018, https://www.washingtonpost.com/news/post-politics /wp/2018/01/02/trump-to-north-korean-leader-kim-my-nuclear-button-is-much-bigger-more -powerful/.

15   "Concern that Trump could launch unjustified nuclear attack," *Washington Post*, January 29, 2018, https://web.archive.org/web/20180129090428/https://www.washingtonpost.com/politics/ polling/concern-trump-could-launch-unjustified-nuclear/2018/01/24/d1c2b920-0034-11e8-86b9 -8908743c79dd_page.html.

16   "No First Use: Summary of Public Opinion," *Rethink Media*, April 19, 2019.

17   Tom Collina, "The Most Dangerous Man in the World," *Defense One*, November 14, 2017, https://www.defenseone.com/ideas/2017/11/most-dangerous-man-world/142542.

18   US Senate, Senate Committee on Foreign Relations, "Corker Statement at Hearing on Authority to Order the Use of Nuclear Weapons," US Senate Press Release, November 14, 2017, https:// www.foreign.senate.gov/press/chair/release/corker-statement-at-hearing-on-authority-to-order-the -use-of-nuclear-weapons.

19   US Senate, Senate Committee on Foreign Relations, "Cardin Remarks at Hearing on Use of Nuclear Weapons," US Senate Press Release, November 14, 2017, https://www.foreign.senate.gov/ press/ranking/release/cardin-remarks-at-hearing-on-use-of-nuclear-weapons.

20   Michael S. Rosenwald, "What if the president ordering a nuclear attack isn't sane? An Air Force major lost his job for asking," *Washington Post*, August 10, 2017, https://www.washingtonpost .com/news/retropolis/wp/2017/08/09/what-if-the-president-ordering-a-nuclear-attack-isnt-sane-a -major-lost-his-job-for-asking.

21   Interview with the authors, October 17, 2019.

22   Ben Cohen, "President Trump, North Korea and Nuclear War," *Jurist*, August 14, 2017, https:// www.jurist.org/commentary/2017/08/president-trump-nuclear-war.

23   Kennette Benedict, *10 Big Nuclear Ideas for the Next President: Add Democracy to Nuclear Policy* (Washington, DC: Ploughshares Fund, 2016).

24   US Department of State, "Memorandum of Discussion at the 214th Meeting of the National Security Council, Denver, September 12, 1954," *Foreign Relations of the United States 1952–54*, vol. 14, 618, quoted in Elaine Scarry, *Thermonuclear Monarchy* (New York: W. W. Norton & Co., 2013), 38.

25   Steven Kull, Nancy Gallagher, Evan Fehsenfeld, Evan Charles Lewitus, and Emmaly Read, *Americans on Nuclear Weapons* (College Park, MD: University of Maryland, May 2019), http:// www.publicconsultation.org/wp-content/uploads/2019/05/Nuclear_Weapons_Report_0519.pdf (accessed June 20, 2019).

26   US House of Representatives, Office of Representative Ted Lieu, "Rep. Lieu and Sen. Markey

Reintroduce Bill to Limit President's Ability to Launch Nuclear First Strike," US House of Representatives press release, January 29, 2019, https://lieu.house.gov/media-center/press-releases/rep-lieu-and-sen-markey-reintroduce-bill-limit-president-s-ability.

27  Senator Mark Warner, Twitter post, December 20, 2018, 5:52 p.m., https://twitter.com/Mark Warner/status/1075886699417403393.

28  Karoun Demirjian and Greg Jaffe, "'A sad day for America': Washington fears a Trump unchecked by Mattis," *Washington Post*, December 20, 2018, https://www.washingtonpost.com/world/national-security/a-sad-day-for-america-washington-fears-a-trump-unchecked-by-mattis/2018/12/20/faef8da0-04ac-11e9-b6a9-0aa5c2fcc9e4_story.html.

29  Bruce Blair and Jon Wolfsthal, "Trump can launch nuclear weapons whenever he wants, with or without Mattis," *Washington Post*, December 23, 2018, https://www.washingtonpost.com/outlook/2018/12/23/trump-can-launch-nuclear-weapons-whenever-he-wants-with-or-without-mattis/?.

30  Nahal Toosi, "'A moment of crisis': Warren lays out foreign policy vision," *Politico*, November 29, 2018, https://www.politico.com/story/2018/11/29/elizabeth-warren-foreign-policy-1029102.

31  US Senate, Office of Senator Elizabeth Warren, "Senator Warren, Chairman Smith Unveil Legislation to Establish 'No-First-Use' Nuclear Weapons Policy," US Senate press release, January 30, 2019, https://www.warren.senate.gov/newsroom/press-releases/senator-warren-chairman-smith-unveil-legislation-to-establish-no-first-use-nuclear-weapons-policy.

32  Mary B. DeRosa and Ashley Nicolas, "The President and Nuclear Weapons: Authorities, Limits, and Process," Nuclear Threat Initiative, December 2019.

33  Alex Wellerstein, "The Kyoto Misconception: What Truman Knew, and Didn't Know, About Hiroshima," in Michael D. Gordin and G. John Ikenberry, eds., *The Age of Hiroshima* (Princeton, NJ: Princeton University Press, 2020), chapter 3.

34  William Burr, "The Atomic Bomb and the End of World War II," George Washington University, The National Security Archive, August 5, 2005, https://nsarchive2.gwu.edu/nukevault/ebb525-The-Atomic-Bomb-and-the-End-of-World-War-II/documents/082.pdf (accessed June 20, 2019).

35  Daniel Ford, *The Button: The Nuclear Trigger—Does It Work?* (London: Unwin Paperbacks, 1986), 34.

36  Burr, "The Atomic Bomb and the End of World War II."

37  Harry Truman, "Statement by the President Reviewing Two Years of Experience with the Atomic Energy Act," Washington, DC: Harry S. Truman Library & Museum, July 24, 1948, https://www.trumanlibrary.gov/library/public-papers/164/statement-president-reviewing-two-years-experience-atomic-energy-act (accessed July 17, 2019).

38  David E. Lilienthal, *Journals of David E. Lilienthal, Volume II: The Atomic Energy Years, 1945–1950* (New York: Harper & Row, 1964), 390–91.

39  Alex Wellerstein, "The Kyoto Misconception."

40  McGeorge Bundy, *Danger and Survival: Choices about the Bomb in the First Fifty Years* (New York: Random House Inc., 2018), 384.

41  US Department of Defense, *Report on Nuclear Employment Strategy of the United States*, June 12, 2013, https://www.globalsecurity.org/wmd/library/policy/dod/us-nuclear-employment-strategy.pdf (accessed June 20, 2019).

42  "Memorandum of a Conversation in the White House between John F. Kennedy, Per Haekkerup, and Others, December 4, 1962," cited in William Burr, "Presidential Control of Nuclear Weapons: The 'Football,'" *Electronic Briefing Book* 632 (July 9, 2018), National Security Archive, https://nsarchive.gwu.edu/briefing-book/nuclear-vault/2018-07-09/presidential-control-nuclear-weapons-football.

43    Mark Hertling, "Nuclear codes: A president's awesome power," CNN, June 10, 2016, https://www .cnn.com/2016/06/09/opinions/nuclear-codes-hertling.

44    Michael Dobbs, "The Real Story of the 'Football' That Follows the President Everywhere," *The Smithsonian*, October 2014, https://www.smithsonianmag.com/history/real-story-football-follows -president-everywhere-180952779.

45    William Manchester, *The Death of a President: November 20–November 25, 1963* (New York: Harper and Row, 1967), 62–63, 261, 321.

46    Dobbs, "The Real Story of the 'Football.'"

47    Steve Fetter, Lisbeth Gronlund, and David Wright, "How to limit presidential authority to order the use of nuclear weapons," *Bulletin of the Atomic Scientists*, January 23, 2018, https://thebulletin .org/2018/01/how-to-limit-presidential-authority-to-order-the-use-of-nuclear-weapons.

48    George Bush, *All the Best, George Bush: My Life in Letters and Other Writings* (New York: Touchstone, 1999), 539, quoted in David E. Hoffman, *The Dead Hand: The Untold Story of the Cold War Arms Race and Its Dangerous Legacy* (New York: Doubleday, 2009), 383.

49    Garrett M. Graff, *Raven Rock: The Story of the U.S. Government's Secret Plan to Save Itself—While the Rest of Us Die* (New York: Simon & Schuster, 2017), 320.

50    "Bill Clinton–Boris Yeltsin Discussions of the Nuclear Football," George Washington University, The National Security Archive, July 9, 2018, https://nsarchive.gwu.edu/briefing-book/nuclear -vault/2018-07-09/presidential-control-nuclear-weapons-football (accessed June 20, 2019).

51    Ibid.

52    Blair and Wolfsthal, "Trump can launch nuclear weapons."

53    SAC, "History of Strategic Air Command, January–June 1968," vol. 1, 301, quoted in William Burr, "Nixon Administration, the 'Horror Strategy,' and the Search for Limited Nuclear Options, 1969–1972," *Journal of Cold War Studies* 7, no. 3 (Summer 2005): 4, https://doi .org/10.1162/1520397054377188 (accessed June 24, 2019).

54    *Annual Report to the Congress, FY 1990*, January 1989, 37, quoted in George Lee Butler, *Uncommon Cause: A Life at Odds with Convention, Volume II: The Transformative Years* (Denver, CO: Outskirts Press, 2016), 11.

55    Curtis E. LeMay, *America Is in Danger* (New York: Funk & Wagnalls, 1968), 82–83, quoted in Bundy, *Danger and Survival*, 321.

56    Butler, *Uncommon Cause*, 273.

57    William Burr, "Launch on Warning: The Development of US Capabilities, 1959–1979," George Washington University, The National Security Archive, April 2001, https://nsarchive2.gwu.edu/ NSAEBB/NSAEBB43 (accessed June 24, 2019).

58    Butler, *Uncommon Cause*, 273.

59    "Excerpts from Bush's Remarks on National Security and Arms Policy," *New York Times*, May 24, 2000, https://www.nytimes.com/2000/05/24/us/2000-campaign-excerpts-bush-s-remarks-national -security-arms-policy.html.

## CHAPTER 2

1    Stephen Ambrose, *Eisenhower, Volume II: The President* (New York: Touchstone Books, 1984), 150, quoted in McGeorge Bundy, *Danger and Survival: Choices about the Bomb in the First Fifty Years* (New York: Random House Inc., 2018), 374.

2    Robert S. McNamara, *Blundering into Disaster* (New York: Pantheon Books, 1986), 8.

3    John G. Hines, Ellis M. Mishulovich, and John F. Shull, *Soviet Intentions 1965–1985, Volume II: Soviet Post-Cold War Testimonial Evidence*, 24, 124, quoted in Svetlana Savranskaya and William

Burr, *Previously Classified Interviews with Former Soviet Officials Reveal US Strategic Intelligence Failure Over Decades* (Washington, DC: George Washington University, 2009).

4    Joshua Coupe et al., "Nuclear Winter Responses to Nuclear War Between the United States and Russia in the Whole Atmosphere Community Climate Model Version 4 and the Goddard Institute for Space Studies Model-E," *Journal of Geophysical Research: Atmospheres* (2019), AGU Online Library, https://agupubs.onlinelibrary.wiley.com/doi/full/10.1029/2019JD030509#accessDenialLayout.

5    Bundy, *Danger and Survival*, 304.

6    Robert M. Gates, *From the Shadows: The Ultimate Insider's Story of Five Presidents and How They Won the Cold War* (New York: Simon & Schuster, 1996), 108.

7    George Lee Butler, *Uncommon Cause: A Life at Odds with Convention, Volume II: The Transformative Years* (Denver, CO: Outskirts Press, 2016), 408.

8    David E. Hoffman, *The Dead Hand: The Untold Story of the Cold War Arms Race and Its Dangerous Legacy* (New York: Doubleday, 2009), 475.

9    Eric Schlosser, *Command and Control* (New York: Penguin Group, 2013), 83, 84.

10   Ibid., 277.

11   McNamara, *Blundering into Disaster*, 99.

12   Bundy, *Danger and Survival*, 448.

13   Scott D. Sagan, "SIOP-62: The Nuclear War Plan Briefing to President Kennedy," *International Security* 12, no. 1 (1987): 22–51, doi:10.2307/2538916, https://www.belfercenter.org/sites/default/files/legacy/files/CMC50/ScottSaganSIOP62TheNuclearWarPlanBriefingtoPresidentKennedy InternationalSecurity.pdf (accessed June 21, 2019).

14   Ibid.

15   Richard Dean Burns and Joseph M. Siracusa, *Historical Dictionary of the Kennedy-Johnson Era* (Lanham, MD: Scarecrow Press, 2007), 36.

16   The White House, "Memorandum of Conversation, Tuesday, July 27, 1976" (Washington, DC: Gerald R. Ford Presidential Library, July 27, 1976), https://www.fordlibrarymuseum.gov/library/document/0314/1553517.pdf (accessed June 13, 2019).

17   Moorer quoted in Meeting of the General Advisory Committee, June 8, 1971, "SALT, Nuclear Testing and Nuclear Strategy," p. 140, FOIA release, copy at National Security Archive, quoted in William Burr, "Nixon Administration, the 'Horror Strategy,' and the Search for Limited Nuclear Options, 1969–1972," *Journal of Cold War Studies* 7, no. 3 (Summer 2005): 65, https://doi.org/10.1162/1520397054377188 (accessed June 21, 2019).

18   Thomas Power, *Design for Survival* (New York: Coward-McCann, 1965), 80–81, quoted in Bundy, *Danger and Survival*, 548.

19   US Senate, Senate Committee on Armed Services, Ninety-sixth Congress, Hearing, First Session, "SALT II Treaty," part II, 779–80, quoted in Bundy, *Danger and Survival*, 548.

20   Robert S. McNamara, *The Essence of Security* (London: Hodder & Stoughton, 1968), 52–53.

21   Bundy, *Danger and Survival*, 544.

22   Hans M. Kristensen and Robert S. Norris, "Global Nuclear Weapons Inventories, 1945–2013," Taylor & Francis Online, November 27, 2015, https://thebulletin.org/2013/10/global-nuclear-weapons-inventories-1945-2013, https://www.tandfonline.com/doi/full/10.1177/0096340213501363 (accessed June 13, 2019).

23   Chalmers Roberts, *First Rough Draft: A Journalist's Journal of Our Times* (New York: Henry Holt & Company, Inc., 1973), quoted in Bundy, *Danger and Survival*, 335.

24   Joseph Alsop, "After Ike, the Deluge," *Washington Post*, October 7, 1959, A17; Stewart Alsop, "Our Gamble with Destiny," *Saturday Evening Post*, May 16, 1959, 23, 114–18; Edgar M. Bottome,

*Missile Gap* (Madison, NJ: Fairleigh Dickinson University Press, 1971), 97, quoted in Bundy, *Danger and Survival*, 338.

25   John Fitzgerald Kennedy, "United States Military and Diplomatic Policies—Preparing for the Gap," Papers of John F. Kennedy, Pre-Presidential Papers, Senate Files, Speeches and the Press, Speech Files, 1953–1960. US Military Power, Senate Floor, August 14, 1958, 6. JFK-SEN-0901-022 (Boston: John F. Kennedy Presidential Library and Museum), https://www.jfklibrary.org/asset-viewer/archives/JFKSEN/0901/JFKSEN-0901-022.

26   Henry Kissinger, *The Necessity for Choice* (New York: HarperCollins, 1961), 26, quoted in Bundy, *Danger and Survival*, 348.

27   Bundy, *Danger and Survival*, 348.

28   Stephen Budiansky, *Code Warriors* (New York: Alfred A. Knopf, 2016), 238–39.

29   McNamara, *Blundering into Disaster*, 26–27.

30   Bundy, *Danger and Survival*, 342.

31   Paul H. Nitze, "Assuring Strategic Stability in an Era of Detente," *Foreign Affairs*, January 1, 1976, https://www.foreignaffairs.com/articles/united-states/1976-01-01/assuring-strategic-stability-era-d-tente.

32   "Intelligence Community Experiment in Competitive Analysis: Soviet Strategic Objectives, an Alternative View: Report of Team B," December 1976, DNSA, SE00501.

33   "Soviet Forces for Strategic Nuclear Conflict Through the Mid-1980s," NIE 11-⅜-76, December 21, 1976, vol. 1, Key Judgments and Summary, 3.

34   Hoffman, *Dead Hand*, 60.

35   McNamara, *Blundering into Disaster*, 48.

36   Ibid., 49.

37   Ibid., 137–38.

38   Ibid., 44.

39   Ibid., 271.

40   Hoffman, *Dead Hand*, 96.

41   David Holloway, *Stalin & the Bomb* (New Haven, CT: Yale University Press, 1994), 1.

42   Malyshev to Khrushchev, "Opasnosti Atomnoi Voiny I Predlozhenie Prezidenta Eizenkhauera," April 1, 1954, TsKhSA f. 5, op. 30, d. 16, pp. 38–44, quoted in ibid., 338.

43   Mohammed Heikal, *Sphinx and Commissar: The Rise and Fall of Soviet Influence in the Arab World* (London: Collins, 1978), 129, quoted in Holloway, *Stalin & The Bomb*, 339.

44   "Rech' N.S. Khrushcheva," *Pravda*, November 28, 195, 1, quoted in ibid., 343.

45   Holloway, *Stalin & The Bomb*, 344.

46   Ibid., 345.

47   Hoffman, *Dead Hand*, 475.

48   Ibid., 253.

49   Interview with the authors, May 3, 2019.

50   US Senate, Senate Committee on Armed Services, "Hearing to Receive Testimony on Nuclear Policy and Posture," Stenographic Transcript, February 28, 2019, https://www.armed-services.senate.gov/imo/media/doc/19-18_02-28-19.pdf.

## CHAPTER 3

1   Interview with the authors, July 10, 2019.

2   Madison Park, "Here's what went wrong with the Hawaii false alarm," CNN, January 31, 2018, https://www.cnn.com/2018/01/31/us/hawaii-false-alarm-investigation-findings/index.html.

# NOTES

3    Cynthia Lazaroff, "Dawn of a new Armageddon," *Bulletin of the Atomic Scientists*, August 6, 2018, https://thebulletin.org/2018/08/dawn-of-a-new-armageddon.

4    "Rep. Tulsi Gabbard Addresses False Hawaii Missile Alert, Calls for Accountability & Immediate Action," YouTube video, 6:01, posted by "Congresswoman Tulsi Gabbard," January 14, 2018, https://www.youtube.com/watch?v=OoqEnpclffg&feature=youtu.be.

5    Interview with the authors, May 15, 2019.

6    David E. Hoffman, *The Dead Hand: The Untold Story of the Cold War Arms Race and Its Dangerous Legacy* (New York: Doubleday, 2009), 475.

7    Ibid., 36–37.

8    Robert S. McNamara, *Blundering into Disaster* (New York: Pantheon Books, 1986), 14.

9    Alex Wellerstein, *The Kyoto Misconception: What Truman Knew, and Didn't Know, About Hiroshima*, 2.

10   McNamara, *Blundering into Disaster*, 10.

11   George Lee Butler, *Uncommon Cause: A Life at Odds with Convention, Volume II: The Transformative Years* (Denver, CO: Outskirts Press, 2016), 409.

12   Daniel Ellsberg, *The Doomsday Machine: Confessions of a Nuclear War Planner* (New York: Bloomsbury Publishing, 2017), 210.

13   Ibid., 221.

14   James A. Winnefeld, Jr., "A Common Policy for Avoiding a Disastrous Nuclear Decision," Carnegie Endowment for International Peace, September 10, 2019, https://carnegieendowment .org/2019/09/10/commonsense-policy-for-avoiding-disastrous-nuclear-decision-pub-79799.

15   Interview with the authors, May 3, 2019.

16   Lawrence K. Altman and Todd Purdum, "In J.F.K. File, Hidden Illness, Pain and Pills," *New York Times*, November 17, 2002, https://www.nytimes.com/2002/11/17/us/in-jfk-file-hidden-illness -pain-and-pills.html.

17   Ed Pilkington, "Ronald Reagan had Alzheimer's while president, says son," *The Guardian*, January 17, 2011, https://www.theguardian.com/world/2011/jan/17/ronald-reagan-alzheimers-president-son.

18   Michael S. Rosenwald, "The US did nothing after North Korea shot down a Navy spy plane in 1969. Trump vows that won't happen again," *Washington Post*, November 7, 2017, https://www .washingtonpost.com/news/retropolis/wp/2017/11/07/north-korea-shot-down-a-u-s-spy-plane-in -1969-trump-might-be-appalled-by-the-response.

19   Anthony Summers and Robbyn Swan, "Drunk in Charge: Part Two," *The Guardian*, September 2, 2000, https://www.theguardian.com/books/extracts/story/0,6761,362959,00.html.

20   Tim Weiner, "That Time the Middle East Exploded—and Nixon Was Drunk," *Politico*, June 15, 2015, https://www.politico.com/magazine/story/2015/06/richard-nixon-watergate-drunk-yom -kippur-war-119021.

21   James M. Acton and Nick Blanchette, "The United States' Nuclear and Non-Nuclear Weapons Are Dangerously Entangled," *Foreign Policy*, November 12, 2019, https://foreignpolicy .com/2019/11/12/the-united-states-nuclear-and-non-nuclear-weapons-are-dangerously-entangled/.

22   Laicie Heeley, "Nixon's Drunken Run-Ins with the Bomb," *Inkstick*, November 13, 2017, https:// inkstickmedia.com/nixons-drunken-run-ins-bomb.

23   Interview with the authors, July 10, 2019.

24   Hoffman, *Dead Hand*, 369.

25   Ellsberg, *Doomsday Machine*, 43.

26   Ibid., 44. Also found in Eric Schlosser, *Command and Control* (New York: Penguin Group, 2013), 255.

27   Richard Halloran, "Nuclear Missiles: Warning System and the Question of When to Fire," *New*

*York Times*, May 29, 1983, https://www.nytimes.com/1983/05/29/us/nuclear-missiles-warning
-system-and-the-question-of-when-to-fire.html.

28    US Department of Defense, *Letter to General Lew Allen Jr.*, by Lt. General James V. Hartinger (Washington, DC: March 14, 1980), https://nsarchive2.gwu.edu/nukevault/ebb371/docs/doc 11.pdf.

29    US Department of State, *Brezhnev Message to President on Nuclear False Attack Alarm*, reviewed by
Theodore Sellin (Washington, DC: November 13, 1985), https://nsarchive2.gwu.edu/nukevault/
ebb371/docs/doc%202%2011-14-79.PDF.

30    Robert M. Gates, *From the Shadows: The Ultimate Insider's Story of Five Presidents and How They
Won the Cold War* (New York: Simon & Schuster, 1996), 114. Scott D. Sagan, *The Limits of Safety:
Organizations, Accidents, and Nuclear Weapons* (Princeton, NJ: Princeton University Press, 1995),
228–29, 238.

31    US Department of State, *Memorandum for the President on Nuclear False Alerts*, Zbigniew Brzezinski (Washington, DC: July 17, 1980), https://nsarchive2.gwu.edu/nukevault/ebb371/docs/doc%
2018%207-12-80.pdf.

32    Ibid.

33    Report of Senator Gary Hart and Senator Barry Goldwater to the Committee on the Armed
Services US Senate, *Recent False Alerts from the Nation's Missile Attack Warning System* (Washington, DC: US Government Printing Office, 1980), 13, https://babel.hathitrust.org/cgi/pt?id=uc1
.31210005931942&view=1up&seq=18.

34    US General Accounting Office, *Report to the Chairman, Committee on Government Operations,
House of Representatives: NORAD's Missile Warning System: What Went Wrong?* (Washington, DC:
May 15, 1981), 4, https://www.gao.gov/assets/140/133240.pdf.

35    David Hoffman, "'I Had a Funny Feeling in My Gut,'" *Washington Post*, February 10, 1999,
http://www.washingtonpost.com/wp-srv/inatl/longterm/coldwar/shatter021099b.htm.

36    *The Man Who Saved the World* is a film based on Petrov, https://www.imdb.com/title/tt2277106.

37    A senior Russian official told Bruce Blair that the alarm, although real, was canceled before it
reached Yeltsin, who then staged his role for the media. Personal communication, July 30, 2019.

38    Union of Concerned Scientists, *Close Calls with Nuclear Weapons*, fact sheet (Cambridge, MA: April
2015), https://www.ucsusa.org/nuclear-weapons/hair-trigger-alert/close-calls#.XGmnLy2ZNQL.

39    Bruce Blair, personal communication, July 30, 2019.

40    Interview with the authors, May 15, 2019.

41    Butler, *Uncommon Cause*, 413.

## CHAPTER 4

1     Interview with the authors, June 19, 2019.

2     Kim Zetter, "An Unprecedented Look at Stuxnet, the World's First Digital Weapon," *Wired*, November 3, 2014, https://www.wired.com/2014/11/countdown-to-zero-day-stuxnet.

3     William Broad, John Markoff, and David Sanger, "Israeli Test on Worm Called Crucial in Iran
Nuclear Delay," *New York Times*, January 15, 2011, https://www.nytimes.com/2011/01/16/world/
middleeast/16stuxnet.html.

4     Ibid.

5     David Sanger, "Obama Order Sped Up Wave of Cyberattacks Against Iran," *New York Times*,
June 1, 2012, https://www.nytimes.com/2012/06/01/world/middleeast/obama-ordered-wave-of
-cyberattacks-against-iran.html.

6     Institute for Science and International Security, *Stuxnet Malware and Natanz: Update of ISIS
December 22, 2010*, report by David Albright, Paul Brannan, and Christina Walrond (Washington,

DC: February 15, 2011), http://isis-online.org/isis-reports/detail/stuxnet-malware-and-natanz -update-of-isis-december-22-2010-reportsupa-href1/8.

7   Sanger, "Obama Order."

8   Ibid.

9   Ralph Langner, "Nitro Zeus Fact Check and Big Picture," February 22, 2016, https://www.langner .com/2016/02/nitro-zeus-fact-check-and-big-picture/.

10  Julian Barnes and Eric Schmitt, "White House Reviews Military Plans Against Iran, in Echoes of Iraq War," *New York Times*, May 13, 2019, https://www.nytimes.com/2019/05/13/world/middle east/us-military-plans-iran.html.

11  Mark Mazzetti and David Sanger, "US Had Cyberattack Plan If Iran Nuclear Dispute Led to Conflict," *New York Times*, February 16, 2016, https://www.nytimes.com/2016/02/17/world/middle east/us-had-cyberattack-planned-if-iran-nuclear-negotiations-failed.html.

12  Nicole Perlroth and David Sanger, "US Escalates Online Attacks on Russia's Power Grid," *New York Times*, June 15, 2019, https://www.nytimes.com/2019/06/15/us/politics/trump-cyber-russia -grid.html.

13  Julian Barnes and David Sanger, "The Urgent Search for a Cyber Silver Bullet Against Iran," *New York Times*, September 23, 2019, https://www.nytimes.com/2019/09/23/world/middleeast/iran -cyberattack-us.html.

14  Mazzetti, and Sanger, "US Had Cyberattack Plan."

15  William Broad and David Sanger, "Trump Inherits a Secret Cyberwar Against North Korean Missiles," *New York Times*, March 4, 2017, https://www.nytimes.com/2017/03/04/world/asia/ north-korea-missile-program-sabotage.html.

16  US Department of Defense, Defense Science Board, *Resilient Military Systems and the Advanced Cyber Threat*, task force report (Washington, DC: US Department of Defense, January 2013), 5, https://apps.dtic.mil/dtic/tr/fulltext/u2/a569975.pdf.

17  Broad and Sanger, "Trump Inherits a Secret Cyberwar."

18  Mikhail Gorbachev, "The Madness of Nuclear Deterrence," *Wall Street Journal*, April 29, 2019, https://www.wsj.com/articles/the-madness-of-nuclear-deterrence-11556577762.

19  Department of Defense, *Resilient Military Systems*, 1.

20  *Nuclear Weapons in the New Cyber Age: Report of the Cyber-Nuclear Weapons Study Group* (Washington, DC: Nuclear Threat Initiative, September 2018), 6, https://media.nti.org/documents/ Cyber_report_finalsmall.pdf.

21  Royal United Services Institute for Defence and Security Studies, *Cyber Threats and Nuclear Weapons: New Questions for Command and Control, Security and Strategy*, occasional paper, Andrew Futter (London: July 2016), 29, https://rusi.org/sites/default/files/cyber_threats_and_nuclear_ combined.1.pdf; Richard J. Danzig, "Surviving on a Diet of Poisoned Fruit: Reducing the National Security Risks of America's Cyber Dependencies," Center for a New American Security, July 2014, 6, http://www.cnas.org/sites/default/files/publications-pdf/CNAS_PoisonedFruit_Danzig_0.pdf.

22  US Department of Defense, Office of the Secretary of Defense, *Nuclear Posture Review* (Washington, DC: February 2018), 57, https://media.defense.gov/2018/Feb/02/2001872886/-1/-1/1/2018 -NUCLEAR-POSTURE-REVIEW-%20FINAL-REPORT.PDF.

23  Department of Defense, *Resilient Military Systems*, 1.

24  Ibid., 42.

25  *Nuclear Weapons in the New Cyber Age*, 19.

26  US Department of Defense, Strategic Command, *US Strategic Command and US Northern Command SASC Testimony*, transcript of General John Hyten as delivered on February 26, 2019

# NOTES

(Washington, DC: March 1, 2019), https://www.stratcom.mil/Media/Speeches/Article/1771903/us-strategic-command-and-us-northern-command-sasc-testimony.

27  Nicole Perlroth, "A Cyberattack 'the World Isn't Ready For,'" *New York Times*, June 22, 2017, https://www.nytimes.com/2017/06/22/technology/ransomware-attack-nsa-cyberweapons.html.

28  Nicole Perlroth, "As Trump and Kim Met, North Korean Hackers Hit Over 100 Targets in US and Ally Nations," *New York Times*, March 3, 2019, https://www.nytimes.com/2019/03/03/technology/north-korea-hackers-trump.html.

29  "Air Force Loses Contact with 50 ICBMs at Wyoming Base," *Nuclear Threat Initiative*, October 27, 2010, https://www.nti.org/gsn/article/air-force-loses-contact-with-50-icbms-at-wyoming-base.

30  Bruce Blair, "Could Terrorists Launch America's Nuclear Missiles?," *TIME*, November 11, 2010, http://content.time.com/time/nation/article/0,8599,2030685,00.html.

31  Bruce Blair, "Why Our Nuclear Weapons Can Be Hacked," *New York Times*, March 14, 2017, https://www.nytimes.com/2017/03/14/opinion/why-our-nuclear-weapons-can-be-hacked.html.

32  US Senate, Committee on Armed Services, *Hearing to Receive Testimony on the US Strategic Command and US Cyber Command in Review of the Defense Authorization Request for Fiscal Year 2014 & Future Years Defense Program* (Washington, DC: March 12, 2013), https://www.hsdl.org/?view&did=733855.

33  Eric Schlosser, "Neglecting Our Nukes," *Politico*, September 16, 2013, https://www.politico.com/story/2013/09/neglecting-our-nukes-096854.

34  US Government Accountability Office, *Weapon Systems Cybersecurity: DoD Just Beginning to Grapple with Scale of Vulnerabilities* (Washington, DC: October 2018), https://www.gao.gov/assets/700/694913.pdf.

35  Department of Defense Office of Inspector General, "Security Controls at DoD Facilities for Protecting Ballistic Missile Defense System Technical Information DODIG-2019-034," December 10, 2018, https://www.dodig.mil/reports.html/Article/1713611/security-controls-at-dod-facilities-for-protecting-ballistic-missile-defense-sy/.

36  William Broad and David Sanger, "New US Weapons Systems Are a Hackers' Bonanza, Investigators Find," *New York Times*, October 10, 2018, https://www.nytimes.com/2018/10/10/us/politics/hackers-pentagon-weapons-systems.html.

37  Patrick Tucker, "Hacking into Future Nuclear Weapons: The US Military's Next Worry," *Defense One*, December 29, 2016, https://www.defenseone.com/technology/2016/12/hacking-future-nuclear-weapons-us-militarys-next-worry/134237.

38  US Department of Defense, *US Strategic Command*.

39  US Department of Defense, Defense Science Board, *Task Force on Cyber Deterrence*, report (Washington, DC: US Department of Defense, February 2017), 22, https://www.acq.osd.mil/dsb/reports/2010s/dsb-cyberdeterrencereport_02-28-17_final.pdf.

40  Ibid.

41  US Department of State, *Statement by the United States in Cluster 1: Nuclear Disarmament and Security Assurances*, delivered by Robert Wood (Washington, DC: May 2, 2019), https://www.state.gov/statement-by-the-united-states-in-cluster-1-nuclear-disarmament-and-security-assurances.

42  US Department of Defense, *Nuclear Posture Review*, 38.

43  US Department of Defense, *US Strategic Command*.

44  US Department of Defense, *Task Force on Cyber Deterrence*, 17.

45  Ibid., 18.

46  Global Zero, *Global Zero Commission on Nuclear Risk Reduction: De-Alerting and Stabilizing the*

*World's Nuclear Force Postures*, report (Washington, DC: April 2015), https://www.globalzero.org/wp-content/uploads/2018/09/global_zero_commission_on_nuclear_risk_reduction_report_0.pdf.

47   *Nuclear Weapons in the New Cyber Age: Report of the Cyber-Nuclear Weapons Study Group* (Washington, DC: Nuclear Threat Initiative, September 2018), 23, https://media.nti.org/documents/Cyber_report_finalsmall.pdf.

## CHAPTER 5

1   Andrei Sakharov, "Sakharov on Gorbachev, Bush," *Washington Post*, December 3, 1989, https://www.washingtonpost.com/archive/opinions/1989/12/03/sakharov-on-gorbachev-bush/3f425cc8-1c02-48b4-99d2-0d022683728f/?utm_term=.66fad6c227b0.

2   David Sanger, "Would Donald Trump Ever Use Nuclear Arms First? He Doesn't Seem Sure," *New York Times*, September 27, 2016, https://www.nytimes.com/2016/09/28/us/politics/donald-trump-hillary-clinton-nuclear-policy-cyber.html.

3   Sakharov, "Sakharov on Gorbachev, Bush."

4   George Lee Butler, *Uncommon Cause: A Life at Odds with Convention, Volume II: The Transformative Years* (Denver, CO: Outskirts Press, 2016), 409–10.

5   Interview with the authors, September 20, 2019.

6   Steve Fetter and Jon Wolfsthal, "No First Use and Credible Deterrence," *Journal for Peace and Nuclear Disarmament* 1, no. 1 (2018): 102–14, https://www.tandfonline.com/doi/full/10.1080/25751654.2018.1454257.

7   McGeorge Bundy, *Danger and Survival: Choices about the Bomb in the First Fifty Years* (New York: Random House Inc., 2018), 230.

8   Harry S. Truman, *Public Papers*, 1952–1953, 1124–25, quoted in Bundy, *Danger and Survival*, 234.

9   Ibid., 245.

10   Foreign Relations of the United States, 1952–1954, "Memorandum of Discussion at the 190th Meeting of the National Security Council, Thursday, March 25, 1954," *National Security Affairs, Volume II, Part 1*, Document 114, drafted by Deputy Executive Secretary Gleason on Friday, March 26, 1954 (Washington, DC: Government Printing Office, 1954), https://history.state.gov/historicaldocuments/frus1952-54v02p1/d114.

11   John Foster Dulles, *John Foster Dulles Papers*, April 7, 1954, Princeton University Library, quoted in Bundy, *Danger and Survival*, 254.

12   Dwight Eisenhower, "Atoms For Peace," speech, United Nations General Assembly, December 8, 1953, International Atomic Energy Agency, https://www.iaea.org/about/history/atoms-for-peace-speech.

13   Thomas Nichols, *No Use: Nuclear Weapons and US National Security* (Philadelphia: University of Pennsylvania Press, 2013), 91, https://books.google.com/books?id=Y_klAgAAQBAJ&pg=PA91&lpg=PA91&dq=#v=onepage&q&f=false.

14   George Kistiakowsky, *A Scientist at the White House* (Cambridge, MA: Harvard University Press, 2014), 399–400, quoted in Bundy, *Danger and Survival*, 322.

15   David Alan Rosenberg, "The Origins of Overkill, Nuclear Weapons and American Strategy, 1945–1960," in Steven E. Miller, ed., *Strategy and Nuclear Deterrence* (Princeton, NJ: Princeton University Press, 1984), 113–81, quoted in David E. Hoffman, *The Dead Hand: The Untold Story of the Cold War Arms Race and Its Dangerous Legacy* (New York: Doubleday, 2009), 16.

16   Lawrence D. Weiler, "No First Use: A History," *Bulletin of the Atomic Scientists* 39, no. 2 (1983): 28–34, https://www.tandfonline.com/doi/abs/10.1080/00963402.1983.11458948.

17   Bundy, *Danger and Survival*, 375.

# NOTES

18   Paul F. Diehl, ed., *Through the Straits of Armageddon* (Athens, GA: University of Georgia Press, 1987), ix, quoted in Bundy, *Danger and Survival*, 354.

19   Donald Brennan, "Strategic Alternatives," *New York Times*, May 24, 1971, 31, quoted in Hoffman, *Dead Hand*, 16.

20   Richard Rhodes, "Absolute Power," *New York Times*, March 21, 2014, https://www.nytimes.com/2014/03/23/books/review/thermonuclear-monarchy-by-elaine-scarry.html.

21   Robert S. McNamara, *Blundering into Disaster* (New York: Pantheon Books, 1986), 139.

22   John F. Kennedy, *Public Papers*, 1962, 543, quoted in Bundy, *Danger and Survival*, 322.

23   McNamara, *Blundering into Disaster*, 16, 139.

24   "NSC Meeting—February 19, 1969," National Archives, Nixon Presidential Materials Project, National Security Council Institutional Files, H-109, National Security Council (NSC) Minutes Originals, 1969, 1, quoted in William Burr, "Nixon Administration, the 'Horror Strategy,' and the Search for Limited Nuclear Options, 1969–1972," *Journal of Cold War Studies* 7, no. 3 (Summer 2005): 34, https://doi.org/10.1162/1520397054377188 (accessed July 11, 2019).

25   Burr, "Nixon Administration," 63.

26   Ibid., 48.

27   H. R. Haldeman, *The Haldeman Diaries* (New York: G. P. Putnam's Sons, 1994), 55, quoted in Hoffman, *Dead Hand*, 19–20.

28   Daniel Ellsberg, *The Doomsday Machine: Confessions of a Nuclear War Planner* (New York: Bloomsbury Publishing, 2017), 310.

29   Nina Tannenwald, *The Nuclear Taboo: The United States and the Non-Use of Nuclear Weapons Since 1945* (Cambridge, UK: Cambridge University Press, 2008), 237.

30   Alexander M. Haig Jr., *Inner Circles: How America Changed the World: A Memoir* (New York: Warner Books, 1992), 28.

31   Martin Anderson, *Revolution: The Reagan Legacy* (New York: Harcourt Brace Jovanovich, 1988), 80–83, quoted in Hoffman, *Dead Hand*, 28.

32   Thomas C. Reed, interview, December 4, 2004, quoted in Hoffman, *Dead Hand*, 39.

33   Ibid.

34   Ronald Reagan, *An American Life* (New York: Simon & Schuster, 2011), 586, quoted in Hoffman, *Dead Hand*, 92.

35   Hugh Sidey, "The Gipper Says Goodbye as New Cast Moves Onstage," *TIME*, January 30, 1989, quoted in William Burr, "Reagan's Nuclear War Briefing Declassified," George Washington University, National Security Archive, December 22, 2016, https://nsarchive.gwu.edu/briefing-book/nuclear-vault/2016-12-22/reagans-nuclear-war-briefing-declassified (accessed July 11, 2019).

36   Bob Woodward, *Obama's Wars* (New York: Simon & Schuster, 2011), 11.

37   US Senate, Senate Committee on Foreign Relations, "The History and Lessons of START," committee hearing, testimony by James A. Baker, transcript, May 19, 2010, https://www.foreign.senate.gov/imo/media/doc/BakerTestimony100519p.pdf (accessed October 30, 2019).

38   Tannenwald, *The Nuclear Taboo*, 298.

39   This story is well told in Steve Fetter and Jon Wolfsthal, "No First Use and Credible Deterrence," *Journal for Peace and Nuclear Disarmament* 1, no. 1 (2018): 102–14, https://www.tandfonline.com/doi/full/10.1080/25751654.2018.1454257.

40   Bruce Blair and James Cartwright, "End the First-Use Policy for Nuclear Weapons," *New York Times*, August 14, 2016, https://www.nytimes.com/2016/08/15/opinion/end-the-first-use-policy-for-nuclear-weapons.html.

41   Fetter and Wolfsthal, "No First Use and Credible Deterrence."

NOTES

42 Interview with the authors, May 3, 2019.
43 Joe Biden, "Remarks by the Vice President on Nuclear Security," The White House, January 11, 2017, https://obamawhitehouse.archives.gov/the-press-office/2017/01/12/remarks-vice-president -nuclear-security (accessed July 11, 2019).
44 Interview with the authors, May 3, 2019.
45 Sanger, "Would Donald Trump Ever Use Nuclear Arms First? He Doesn't Seem Sure."
46 Anatoly Antonov, "America, You're Not Listening to Us," *Defense One*, April 7, 2019, https://www .defenseone.com/ideas/2019/04/america-youre-not-listening-us/156110.
47 Connor O'Brien, "Shultz warns Congress against low-yield nuclear weapons," *Politico*, January 25, 2018, https://www.politico.com/story/2018/01/25/nuclear-weapons-george-schultz-369450.
48 Richard Sisk, "Mattis: There Is No Such Thing as a 'Tactical' Nuke," Military.com, February 6, 2018, https://www.military.com/defensetech/2018/02/06/mattis-there-no-such-thing-tactical -nuke.html.
49 Adam Smith, "Keynote with Representative Adam Smith," keynote address with questions and answers, Carnegie International Nuclear Policy Conference 2019, Carnegie Endowment for International Peace, Washington, DC, March 12, 2019, https://carnegieendowment.org/2019/03/12/ keynote-with-representative-adam-smith-pub-78883.
50 McGeorge Bundy, George Kennan, Robert McNamara, Gerard Smith, "Nuclear Weapons and the Atlantic Alliance," *Foreign Affairs*, Spring 1982, https://www.foreignaffairs.com/articles/united -states/1982-03-01/nuclear-weapons-and-atlantic-alliance.
51 Ellsberg, *Doomsday Machine*, 319–322.
52 Press Release, "Feinstein Urges President to Declare No-First-Use Nuclear Policy," August 9, 2016, https://www.feinstein.senate.gov/public/index.cfm/2016/8/feinstein-urges-president-to -declare-no-first-use-nuclear-policy.
53 Steven Kull, Nancy Gallagher, Evan Fehsenfeld, Evan Charles Lewitus, and Emmaly Read, *Americans on Nuclear Weapons* (College Park, MD: University of Maryland, May 2019), http:// www.publicconsultation.org/wp-content/uploads/2019/05/Nuclear _Weapons_Report_0519.pdf (accessed June 20, 2019).

## CHAPTER 6

1 Adam Smith, "The Future of US Nuclear Policy," Speaker Session, Washington, DC, Ploughshares Fund, November 14, 2018, https://www.ploughshares.org/issues-analysis/article/rep-adam -smith-future-us-nuclear-policy (accessed July 11, 2019).
2 Congressional Budget Office, *Approaches for Managing the Costs of US Nuclear Forces, 2017 to 2046, October 2017*. CBO estimates the cost at $1.2 trillion without inflation and $1.7 trillion with inflation. Since then, costs have risen and the Trump administration has added to the program. For the purposes of this discussion, we round up to $2 trillion.
3 James Miller, "The Future of US-Russia Arms Control," panel, Carnegie International Nuclear Policy Conference 2019, Carnegie Endowment for International Peace, Washington, DC, March 12, 2019, https://carnegieendowment.org/2019/03/11/future-of-u.s.-russia-arms-control-pub-78865.
4 US Department of Defense, *Report on Nuclear Employment Strategy of the United States* (Washington, DC: June 12, 2013), 5, https://www.globalsecurity.org/wmd/library/policy/dod/us-nuclear -employment-strategy.pdf.
5 Interview with the authors, May 3, 2019.
6 Interview with the authors, June 13, 2019.
7 Eugene Sevin, "The MX/Peacekeeper and SICBM: A Search for Survivable Basing," *Defense Systems*

*Information Analysis Center (DSIAC)*, Winter 2017, https://www.dsiac.org/resources/journals/dsiac/winter-2017-volume-4-number-1/mxpeacekeeper-and-sicbm-search-survivable.

8    US Department of Defense, Strategic Command, *US Strategic Command and US Northern Command SASC Testimony*, transcript of Gen. John Hyten as delivered on February 26, 2019 (Washington, DC: March 1, 2019), https://www.stratcom.mil/Media/Speeches/Article/1771903/us-strategic-command-and-us-northern-command-sasc-testimony.

9    Transcript of remarks by Representative Adam Smith at the National Press Club, October 24, 2019, https://www.ploughshares.org/issues-analysis/article/rep-adam-smith-us-nuclear-policy.

10   George Wilson, "Fresh Challenge Voiced to Missile 'Shell Game,'" *Washington Post*, July 24, 1978, https://www.washingtonpost.com/archive/politics/1978/07/24/fresh-challenge-voiced-to-missile-shell-game/7b2f8ae3-0109-43e3-a6bd-2bc04c56f6c8.

11   US Senate, Committee on Armed Services, "Statement of James N. Mattis before the Senate Armed Services Committee," Hearing, January 27, 2015, https://www.armed-services.senate.gov/imo/media/doc/Mattis_01-27-15.pdf.

12   Aaron Mehta, "Mattis Enthusiastic on ICBMs, Tepid on Nuclear Cruise Missile," *Defense News*, January 12, 2017, https://www.defensenews.com/space/2017/01/12/mattis-enthusiastic-on-icbms-tepid-on-nuclear-cruise-missile.

13   Adam Lowther, "Making America's ICBMs Great Again," *Defense One*, January 31, 2017, https://www.defenseone.com/ideas/2017/01/making-americas-icbms-great-again/135024/?oref=d-river&&&utm_term=Editorial%20-%20Early%20Bird%20Brief.

14   US Department of Defense, Strategic Command, transcript of Gen. John Hyten.

15   McGeorge Bundy, George Kennan, Robert McNamara, and Gerard Smith, "Nuclear Weapons and the Atlantic Alliance," *Foreign Affairs*, Spring 1982, https://www.foreignaffairs.com/articles/united-states/1982-03-01/nuclear-weapons-and-atlantic-alliance.

16   Miller, "The Future of US-Russia Arms Control."

17   Anatoly Antonov, "The Future of US-Russia Arms Control," panel, Carnegie International Nuclear Policy Conference 2019, Carnegie Endowment for International Peace, Washington, DC, March 12, 2019, https://carnegieendowment.org/2019/03/11/future-of-u.s.-russia-arms-control-pub-78865.

18   Dean Rusk, *As I Saw It* (London: Penguin Publishing, 1991), 252.

19   Christopher Preble, "Can America Still Afford Its Nuclear Triad?," *The CATO Institute*, July 12, 2016, https://www.cato.org/publications/commentary/can-america-still-afford-its-nuclear-triad.

20   Benjamin Friedman, Christopher Preble, and Matt Fay, "The End of Overkill? Reassessing US Nuclear Weapons Policy," *The CATO Institute*, 2013, 9, https://object.cato.org/sites/cato.org/files/pubs/pdf/the_end_of_overkill_wp_web.pdf.

21   The Democratic Platform Committee, "The Democratic Platform—2016 Democratic National Convention," Democratic National Convention, July 8, 2016, https://democrats.org/wp-content/uploads/2018/10/2016_DNC_Platform.pdf.

22   Adam Smith, "Keynote with Representative Adam Smith," keynote address with questions and answers, Carnegie International Nuclear Policy Conference 2019, Carnegie Endowment for International Peace, Washington, DC, March 12, 2019, https://carnegieendowment.org/2019/03/12/keynote-with-representative-adam-smith-pub-78883.

23   Interview with the authors, May 3, 2019.

24   Interview with the authors, May 3, 2019.

25   Interview with the authors, May 3, 2019.

26   Colin Clark, "OMB Plan to Slice SSBN-X Won't Save Dough, DoD Says," *Breaking Defense*,

November 16, 2011, http://breakingdefense.com/2011/11/omb-plan-to-slice-ssbn-x-fleet-wont
-save-dough-dod-says.

27  Tom Collina and William Perry, "The US Does Not Need New Tactical Nukes," *Defense One*,
April 26, 2018, https://www.defenseone.com/ideas/2018/04/us-does-not-need-new-tactical
-nukes/147757.

28  Smith, "Keynote."

29  Hans Kristensen, "The Flawed Push for New Nuclear Weapons Capabilities," Federation of American
Scientists, June 29, 2017, https://fas.org/blogs/security/2017/06/new-nukes.

30  Madelyn Creedon, "A Question of Dollars and Sense: Assessing the 2018 Nuclear Posture Review,"
Arms Control Association, March 2018, https://www.armscontrol.org/print/9279.

31  US House of Representatives, Committee on Armed Services, "House Armed Services Committee
Hearing on Outside Perspective on Nuclear Deterrence, Policy and Posture," testimony by Dr.
Bruce Blair, March 6, 2019.

32  Bruce Blair, Emma Claire Foley, and Jessica Sleight, *The End of Nuclear Warfighting: Moving to a
Deterrence-Only Posture: An Alternative US Nuclear Posture Review* (Washington, DC: Global Zero,
September 2018), 42, https://www.globalzero.org/wp-content/uploads/2018/09/ANPR-Final.pdf.

33  Bruce Blair, interview with the authors, May 15, 2019.

34  Interview with the authors, September 16, 2019.

35  Interview with the authors, May 15, 2019.

36  "Year-by-Year Data Underlying CBO's Estimate of Nuclear Costs," estimated annual spending on
C3I from Congress of the US, Congressional Budget Office, January 24, 2019, https://www.cbo
.gov/publication/54914.

37  "US Nuclear Excess: Understanding the Costs, Risks, and Alternatives," Arms Control Association,
April 2019, https://www.armscontrol.org/reports/2019/USnuclearexcess.

## CHAPTER 7

1  Shervin Taheran, "Select Reactions to the INF Treaty Crisis," Arms Control Association, February
1, 2019, https://www.armscontrol.org/blog/2018/select-reactions-inf-treaty-crisis.

2  Donald Trump, "President Donald J. Trump to Withdraw the United States from the Intermediate-Range Nuclear Forces (INF) Treaty," The White House, February 1, 2019, https://
www.whitehouse.gov/briefings-statements/president-donald-j-trump-withdraw-united-states
-intermediate-range-nuclear-forces-inf-treaty (accessed July 12, 2019).

3  Anton Troianovski, "Following US, Putin suspends nuclear pact and promises new weapons,"
*Washington Post*, February 2, 2019, https://www.washingtonpost.com/world/following-us
-putin-suspends-nuclear-pact-and-promises-new-weapons/2019/02/02/8160c78e-26e3-11e9
-ad53-824486280311_story.html.

4  US Department of Defense, "Report on Nuclear Employment Strategy of the United States," June
12, 2013, http://www.defense.gov/pubs/ReporttoCongressonUSNuclearEmploymentStrategy
_Section491.pdf.

5  Stephen Schwartz, author of *Atomic Audit*, personal communication with the authors, April 5, 2019.

6  Office of the Director of Intelligence of the United States of America, *Director of National Intelligence Daniel Coats on Russia's Intermediate-Range Nuclear Forces (INF) Treaty Violation*, press briefing delivered by Daniel Coats, November 30, 2018, https://www.dni.gov/index.php/newsroom/
speeches-interviews/item/1923-director-of-national-intelligence-daniel-coats-on-russia-s-inf-treaty
-violation.

7  Matt Korda and Hans Kristensen, "Sunday's US Missile Launch, Explained," Federation of Amer-

ican Scientists, August 20, 2019, https://fas.org/blogs/security/2019/08/sundays-us-missile-launch-explained.

8   Sabra Ayres and David Cloud, "Putin's warning on missiles in Europe pushes US and Russia closer to new arms race," *Los Angeles Times*, February 20, 2019, https://www.latimes.com/world/europe/la-fg-russia-putin-20190220-story.html.

9   Tytti Erästö, *Between the Shield and the Sword: NATO's Overlooked Missile Defense Dilemma* (Washington, DC: Ploughshares Fund, June 2017), 14.

10  Ayres and Cloud, "Putin's warning."

11  Anatoly Antonov, "America, You're Not Listening to Us," *Defense One*, April 7, 2019, https://www.defenseone.com/ideas/2019/04/america-youre-not-listening-us/156110.

12  US Senate, Senate Committee on Armed Services, "US Strategic Command and US Northern Command SASC Testimony," testimony, stenographic transcript, March 1, 2019, https://www.stratcom.mil/Media/Speeches/Article/1771903/us-strategic-command-and-us-northern-command-sasc-testimony (accessed October 30, 2019).

13  US House of Representatives, Office of Speaker Pelosi, "Pelosi Statement on Trump Administration Withdrawal from the INF Treaty," press release, February 1, 2019, https://www.speaker.gov/newsroom/2119-2.

14  US Senate, Senate Committee on Armed Services, "Nomination—Selva," confirmation hearing, video, July 18, 2017, https://www.armed-services.senate.gov/hearings/17-07-18-nomination_—selva (accessed July 11, 2019).

15  George Shultz, "We Must Preserve This Nuclear Treaty," *New York Times*, October 25, 2018, https://www.nytimes.com/2018/10/25/opinion/george-shultz-nuclear-treaty.html.

16  Mikhail Gorbachev, "A New Nuclear Arms Race Has Begun," *New York Times*, October 25, 2018, https://www.nytimes.com/2018/10/25/opinion/mikhail-gorbachev-inf-treaty-trump-nuclear-arms.html.

17  John Bolton and Paula DeSutter, "A Cold War Missile Treaty That's Doing Us Harm," *Wall Street Journal*, August 15, 2011, https://www.wsj.com/articles/SB10001424053111903918104576500273389091098.

18  John Bolton and John Yoo, "An Obsolete Nuclear Treaty Even Before Russia Cheated," *Wall Street Journal*, September 4, 2019, https://www.wsj.com/articles/john-bolton-and-john-yoo-an-obsolete-nuclear-treaty-even-before-russia-cheated-1410304847.

19  Jonathan Landay and David Rhode, "Exclusive: In call with Putin, Trump denounced Obama-era nuclear arms treaty—sources," *Reuters*, February 9, 2017, https://www.reuters.com/article/us-usa-trump-putin-idUSKBN15O2A5.

20  Euan McKirdy, "White House: Trump knows what START Treaty is," CNN, February 10, 2017, https://www.cnn.com/2017/02/10/politics/trump-putin-start-treaty.

21  "The Honorable John Bolton LIVE from YAF's 41st NCSC," YouTube video, posted by YAFTV, July 30, 2019, https://www.youtube.com/watch?v=MHOwCV4xFg8.

22  Julian E. Barnes and David E. Sanger, "Russia Deploys Hypersonic Weapon, Potentially Renewing Arms Race," *New York Times*, December 27, 2019, https://www.nytimes.com/2019/12/27/us/politics/russia-hypersonic-weapon.html.

23  Anatoly Antonov, "The Future of US-Russia Arms Control," panel, Carnegie International Nuclear Policy Conference 2019, Carnegie Endowment for International Peace, Washington, DC, March 12, 2019, https://carnegieendowment.org/2019/03/11/future-of-u.s.-russia-arms-control-pub-78865.

24  Mike Eckel, "US Concludes White Sea Radiation Explosion Came During Russian Nuclear-Missile

Recovery," Radio Free Europe/Radio Liberty, October 12, 2019, https://www.rferl.org/a/u-s
-concludes-white-sea-radiation-explosion-came-during-russian-nuclear-missile-recovery/30213494.
html.

25   Antonov, "America, You're Not Listening."

26   Daryl Kimball, "New START Must Be Extended, with or without China," *National Interest*,
May 27, 2019, https://nationalinterest.org/feature/new-start-must-be-extended-or-without-china
-59227.

27   US Senate, Senate Committee on Foreign Relations, "The Future of Arms Control Post-Intermediate-
Range Nuclear Forces Treaty," committee hearing, video, May 15, 2019, https://www.foreign.senate
.gov/hearings/the-future-of-arms-control-post-intermediate-range-nuclear-forces-treaty (accessed
July 11, 2019).

28   Ibid.

29   Interview with the authors, June 13, 2019.

30   US Department of Defense, Strategic Command, *US Strategic Command and US Northern Command SASC Testimony*, transcript of Gen. John Hyten as delivered on February 26, 2019 (Washington, DC: March 1, 2019), https://www.stratcom.mil/Media/Speeches/Article/1771903/us-strategic
-command-and-us-northern-command-sasc-testimony.

31   Includes active and inactive warheads, strategic and tactical. Does not include thousands of retired
warheads awaiting dismantlement. US Department of State, "Increasing Transparency in the US
Nuclear Weapons Stockpile," fact sheet, May 3, 2010, https://fas.org/sgp/othergov/dod/stockpile.pdf.

32   Joshua Coupe et al., "Nuclear Winter Responses to Nuclear War Between the United States and
Russia in the Whole Atmosphere Community Climate Model Version 4 and the Goddard Institute
for Space Studies Model-E," *Journal of Geophysical Research: Atmospheres* (2019), AGU Online
Library, https://agupubs.onlinelibrary.wiley.com/doi/full/10.1029/2019JD030509#accessDenial
Layout.

33   Lidia Kelly, "Russia can turn US to radioactive ash—Kremlin-backed journalist," *Reuters*, March
16, 2014, https://www.reuters.com/article/ukraine-crisis-russia-kiselyov/russia-can-turn-us-to
-radioactive-ash-kremlin-backed-journalist-idUSL6N0MD0P920140316.

34   "Nuclear Reductions Make the United States Safer: Section 2," October 2014, Arms Control
Association, https://www.armscontrol.org/projects-reports/2014-10/section-2-nuclear-reductions
-make-united-states-safer.

35   Federation of American Scientists, "Increasing Transparency in the US Nuclear Weapons Stockpile," fact sheet, May 3, 2010, https://fas.org/sgp/othergov/dod/stockpile.pdf.

36   George W. Bush, "Remarks Following Discussions with Prime Minister Tony Blair of the United
Kingdom and an Exchange with Reporters," transcript, November 7, 2001, http://www.presidency
.ucsb.edu/ws/index.php?pid=63649&st=nuclear&st1=reduction.

37   US Department of Defense, "Report on Nuclear Employment Strategy."

38   Thom Shanker, "Senator Urges Bigger Cuts to Nuclear Arsenal," *New York Times*, June 14, 2012,
http://www.nytimes.com/2012/06/15/us/politics/senator-levin-urges-bigger-cuts-to-nuclear-arsenal
.html.

39   Max Bergmann, "Colin Powell: 'Nuclear Weapons Are Useless,'" Think Progress, January 27, 2010,
http://thinkprogress.org/security/2010/01/27/175869/colin-powell-nuclear-weapons-are-useless.

40   Ministry of Foreign Affairs of Japan, "Statement of the Non-Proliferation and Disarmament
Initiative (NPDI), 8th Ministerial Meeting, Hiroshima," April 12, 2014, https://www.mofa.go.jp/
files/000035199.pdf. The NPDI includes the countries of Australia, Canada, Chile, Germany,

# NOTES

Japan, Mexico, the Netherlands, Nigeria, the Philippines, Poland, Turkey, and the United Arab Emirates.

## CHAPTER 8

1  Dwight Eisenhower, "Atoms for Peace," speech, United Nations General Assembly, December 8, 1953, International Atomic Energy Agency, https://www.iaea.org/about/history/atoms-for-peace -speech.

2  US Missile Defense Agency, "Homeland Missile Defense System Successfully Intercepts ICBM Target," March 25, 2019, https://www.mda.mil/news/19news0003.html.

3  US Senate, Committee on Armed Services, "To Receive Testimony on Missile Defense Policies and Programs in Review of the Defense Authorization Request for Fiscal Year 2020 and the Future Years Defense Program," stenographic transcript, April 3, 2019, https://www.armed-services.senate .gov/imo/media/doc/19-33_04-03-19.pdf.

4  David William, "There's a flaw in the homeland missile defense system. The Pentagon sees no need to fix it," Los Angeles Times, February 26, 2017, https://www.latimes.com/nation/la-na-missile -defense-flaw-20170226-story.html.

5  Ankit Panda and Vipin Narang, "Deadly Overconfidence: Trump Thinks Missile Defenses Work Against North Korea, and That Should Scare You," War on the Rocks, October 16, 2017, https:// warontherocks.com/2017/10/deadly-overconfidence-trump-thinks-missile-defenses-work-against -north-korea-and-that-should-scare-you.

6  CIA Directorate of Intelligence, Different Times, Same Playbook: Moscow's Response to US Plans for Missile Defense, undated.

7  Tom Collina, Daryl Kimball, and ACA Research Staff, The Case for the New Strategic Arms Reduc- tion Treaty (Washington, DC: Arms Control Association, 2010), https://armscontrol.org/system/ files/NewSTART_Report_FINAL_Nov_30.pdf.

8  Treaty on the Limitation of Anti-Ballistic Missile Systems, https://fas.org/nuke/control/abmt/text/ abm2.htm.

9  James Glanz, "Antimissile Test Viewed as Flawed by Its Opponents," New York Times, January 14, 2000, https://www.nytimes.com/2000/01/14/us/antimissile-test-viewed-as-flawed-by-its -opponents.html.

10  Eric Schmitt, "President Decides to Put Off Work on Missile Shield," New York Times, September 2, 2000, https://archive.nytimes.com/www.nytimes.com/library/world/global/090200missile -defense.html.

11  Gary L. Gregg II, "George W. Bush: Foreign Affairs," UVA Miller Center, https://millercenter.org/ president/gwbush/foreign-affairs.

12  Laura Grego and David Wright, "We Can't Count on Missile Defense to Defeat Incoming Nukes," Scientific American Magazine 320, no. 6 (June 2019), https://www.scientificamerican .com/magazine/sa/2019/06-01/?redirect=1.

13  John King, "Bush Rolls Out Missile Defense System," CNN, December 18, 2002, http://edition .cnn.com/2002/US/12/17/bush.missile/index.html.

14  Tom Collina, "A Little Hit and a Big Miss," Foreign Policy, June 30, 2014, https://foreignpolicy .com/2014/06/30/a-little-hit-and-a-big-miss.

15  Barack Obama, "Arms Control Today, 2008 Presidential Q&A: President-elect Barack Obama," Arms Control Association, September 10, 2008, https://www.armscontrol.org/2008election.

16  Andrea Shalal, "US missile defense system hits target in key test," Reuters, June 22, 2014, https://

www.reuters.com/article/us-usa-military-boeing/u-s-missile-defense-system-hits-target-in-key-test-idUSKBN0EX11Y20140622.

17    US Joint Chiefs of Staff, James Winnefeld Jr., "Adm. Winnefeld's remarks at Atlantic Council's US Missile Defense Plans and Priorities Conference," speech transcript, May 28, 2014, https://www.jcs.mil/Media/Speeches/Article/571961/adm-winnefelds-remarks-at-atlantic-councils-us-missile-defense-plans-and-priori.

18    The White House, "Remarks by the President on Strengthening Missile Defense in Europe" (Washington, DC: Office of the Press Secretary, September 17, 2009), https://obamawhitehouse.archives.gov/the-press-office/remarks-president-strengthening-missile-defense-europe.

19    Peter Baker, "White House Scraps Bush's Approach to Missile Shield," *New York Times*, September 17, 2009, https://www.nytimes.com/2009/09/18/world/europe/18shield.html.

20    Ibid.

21    Rachel Oswald, "US Looking 'Very Hard' at Future of Missile Interceptor: Pentagon," *Nuclear Threat Initiative*, March 12, 2013, https://www.nti.org/gsn/article/us-looking-very-hard-halting-development-icbm-interceptor-miller.

22    Tom Collina, "Phasing Out," *Foreign Policy*, March 14, 2013, https://foreignpolicy.com/2013/03/14/phasing-out.

23    Ibid.

24    Tom Collina, "Pentagon Shifts Gears on Missile Defense," Arms Control Association, April 2013, https://www.armscontrol.org/act/2013_04/Pentagon-Shifts-Gears-on-Missile-Defense.

25    *Trump's Dangerous Missile Defense Buildup* 11, no. 2 (Washington, DC: Arms Control Association, January 17, 2019), https://www.armscontrol.org/issue-briefs/2019-01/trumps-dangerous-missile-defense-buildup.

26    *Report of the American Physical Society Study Group on Boost-Phase Intercept Systems for National Missile Defense: Scientific and Technical Issues* (College Park, MD: American Physical Society, October 5, 2004), https://journals.aps.org/rmp/pdf/10.1103/RevModPhys.76.S1.

27    "UCS Satellite Database," Union of Concerned Scientists, https://www.ucsusa.org/nuclear-weapons/space-weapons/satellite-database#.W2Biz9hKii5 (accessed July 11, 2019).

28    National Research Council, *Making Sense of Ballistic Missile Defense: An Assessment of Concepts and Systems for US Boost-Phase Missile Defense in Comparison to Other Alternatives* (Washington, DC: The National Academies Press, 2012), https://doi.org/10.17226/13189.

29    Laura Grego, David Wright, and Stephen Young, "Space-Based Missile Defense," Union of Concerned Scientists, fact sheet, May 2011, https://www.ucsusa.org/sites/default/files/legacy/assets/documents/nwgs/space-based-md-factsheet-5-6-11.pdf (accessed July 11, 2019).

30    Tom Collina and Zack Brown, "Congress Rushes to Spend Billions on Space Weapons—Even If They Don't Work," *Defense One*, August 3, 2018, https://www.defenseone.com/ideas/2018/08/congress-billions-space-weapons/150273.

## CHAPTER 9

1    McGeorge Bundy, *Danger and Survival: Choices about the Bomb in the First Fifty Years* (New York: Random House Inc., 2018), 130.

2    *States Agree to Ban Nuclear Weapons* (Geneva: International Campaign to Abolish Nuclear Weapons, July 7, 2017), https://www.icanw.org/states_agree_to_ban_nuclear_weapons.

3    Ibid.

4    Interview with the authors, April 30, 2019.

# NOTES

5   *Humanitarian Pledge: Stigmatize, prohibit and eliminate nuclear weapons* (Geneva: International Campaign to Abolish Nuclear Weapons, 2017), http://www.icanw.org/pledge.

6   Steven Pifer, "10 years after Obama's nuclear-free vision, the US and Russia head in the opposite direction," Brookings Institute, April 4, 2019, https://www.brookings.edu/blog/order-from-chaos/2019/04/04/10-years-after-obamas-nuclear-free-vision-the-us-and-russia-head-in-the-opposite-direction.

7   Mikhail Gorbachev, "The Madness of Nuclear Deterrence," *Wall Street Journal*, April 29, 2019, https://www.wsj.com/articles/the-madness-of-nuclear-deterrence-11556577762.

8   "Report of the Committee on Political and Social Problems," Manhattan Project Metallurgical Laboratory, University of Chicago, June 11, 1945 (Franck report), quoted in Joseph Cirincione, *Bomb Scare: The History & Future of Nuclear Weapons* (New York: Columbia University Press, 2007), 16.

9   Ibid., 15.

10  Aage Bohr, "The War Years and the Prospects Raised by the Atomic Weapons," Oppenheimer, "Niels Bohr and Atomic Weapons," 6, quoted in Bundy, *Danger and Survival*, 114.

11  Science Panel's Report to the Interim Committee, June 16, 1945, http://www.atomicarchive.com/Docs/ManhattanProject/Interim.shtml.

12  Harry Truman, "Conversation on the Existence of the Bomb, July 24, 1954," letter, Nuclear Age Peace Foundation, *Foreign Relations of the United States: Conference of Berlin* 1, no. 378, http://www.nuclearfiles.org/menu/library/correspondence/truman-harry/corr_truman_1945-07-24.htm (accessed July 16, 2019).

13  Bundy, *Danger and Survival*, 25–26.

14  Smuts to Churchill, June 15, 1944, quoted in Bundy, *Danger and Survival*, 125.

15  Bundy, *Danger and Survival*, 126.

16  Harry Truman, *Off the Record* 56, *Public Papers*, 1945, 213, quoted in Bundy, *Danger and Survival*, 133.

17  *New York Times*, September 22, 1945, 3, quoted in Bundy, *Danger and Survival*, 140.

18  Acheson's memo of September 25 is in *Foreign Relations of the United States*, vol. 2 (1945): 48–50, quoted in Bundy, *Danger and Survival*, 141.

19  Truman, *Public Papers*, 1945, 379–84, quoted in Bundy, *Danger and Survival*, 142–43.

20  Public Papers of the Presidents of the United States: Harry S. Truman, 1945, Volume 1, 164.

21  Oscar Anderson and Richard Hewlett, *The New World, 1939/1946: A History of the United States Atomic Energy Commission*, vol. 1 (University Park, PA: Pennsylvania State University Press, 1962), 555, quoted in Bundy, *Danger and Survival*, 162.

22  David Holloway, *Entering the Nuclear Arms Race*, 183, quoted in Bundy, *Danger and Survival*, 177.

23  *Foreign Relations of the United States*, vol. 1, 1946, 861–65, quoted in Bundy, *Danger and Survival*, 185.

24  Parliament of Great Britain, "Churchill's speech of March 1, 1955," *Hansard's Parliamentary Debates* (House of Commons), 5th series, 537, quoted in Bundy, *Danger and Survival*, 198.

25  Herbert York, *The Advisors* (Stanford, CA: Stanford University Press, 1989), 155–56, quoted in Bundy, *Danger and Survival*, 208.

26  Brien McMahon, "Letter to President Truman, 1949," in *Foreign Relations of the United States*, vol. 1, 30–31, quoted in Bundy, *Danger and Survival*, 211.

27  David Lilienthal, *The Journals*, vol. 2, 632–33, quoted in Bundy, *Danger and Survival*, 213.

28  Eric Schlosser, *Command and Control* (New York: Penguin Group, 2013), 124.

29  York, *The Advisors*, 155–56, quoted in Bundy, *Danger and Survival*, 214.

30    Bundy, *Danger and Survival*, 217.

31    Andrei Sakharov, *Memoirs*, translated from the Russian by Richard Lourie (New York: Knopf, 1990).

32    David Alan Rosenberg, "The Origins of Overkill, Nuclear Weapons and American Strategy, 1945–1960," 23, quoted in Bundy, *Danger and Survival*, 231.

33    Robert S. McNamara, *Blundering into Disaster* (New York: Pantheon Books, 1986), 55.

34    Ibid., 56.

35    Ibid., 57.

36    Ibid., 58.

37    Henry Kissinger, "Memorandum for the President: Analysis of Strategic Arms Limitation Proposals," George Washington University, The National Security Archives, May 23, 1969, https://nsarchive2.gwu.edu/NSAEBB/NSAEBB60/abm01.pdf (accessed July 17, 2019).

38    William Burr, "The Secret History of the ABM Treaty, 1969–1972," George Washington University, The National Security Archives, November 8, 2001, https://nsarchive2.gwu.edu/NSAEBB/NSAEBB60 (accessed July 17, 2019).

39    Michael Krepon, "Retrospectives on MIRVing in the First Nuclear Age," *Arms Control Wonk*, April 5, 2016, https://www.armscontrolwonk.com/archive/1201264/retrospectives-on-mirving-in-the-first-nuclear-age.

40    Walter Isaacson, *Kissinger: A Biography* (New York: Simon & Schuster, 2005), 325.

41    Ronald Reagan, *Presidential Documents*, June 23, 1986, 839, quoted in Bundy, *Danger and Survival*, 572.

42    Philip Taubman, *The Partnership: Five Cold Warriors and Their Quest to Ban The Bomb* (New York: Harper, 2012).

43    "Record of Conversation between Mikhail Gorbachev and James Baker," George Washington University, The National Security Archives, February 9, 1990, https://nsarchive2.gwu.edu/dc.html?doc=4325680-Document-06-Record-of-conversation-between (accessed July 17, 2019).

44    David Majumdar, "Newly Declassified Documents: Gorbachev Told NATO Wouldn't Move Past East German Border," *The National Interest*, December 12, 2017, https://nationalinterest.org/blog/the-buzz/newly-declassified-documents-gorbachev-told-nato-wouldnt-23629.

45    William Perry, *My Journey at the Nuclear Brink* (Stanford, CA: Stanford University Press, 2015), 118.

46    Ibid., 123.

47    Zoltan Barany, *The Future of NATO Expansion: Four Case Studies* (Cambridge, UK: Cambridge University Press, 2003).

48    Thomas Friedman, "World Affairs; Now a Word from X," *New York Times*, May 2, 1998, quoted in Perry, *My Journey*, 147.

49    George Shultz, William Perry, and Sam Nunn, "The Threat of Nuclear War Is Still with Us," *Wall Street Journal*, April 10, 2019, https://www.wsj.com/articles/the-threat-of-nuclear-war-is-still-with-us-11554936842?.

50    The Senate passed the START II Treaty in January 1996, but this treaty was signed by President George H. W. Bush and never went into force. Russia withdrew from START II after President George W. Bush withdrew from the ABM Treaty in 2002.

51    "Senate Rejects Comprehensive Test Ban Treaty; Clinton Vows to Continue Moratorium," Arms Control Association, July 1999, https://www.armscontrol.org/act/1999_09-10/ctbso99.

52    "Nuclear Testing and Comprehensive Test Ban Treaty (CTBT) Timeline," Arms Control Association, June 2019, https://www.armscontrol.org/factsheets/Nuclear-Testing-and-Comprehensive-Test-Ban-Treaty-CTBT-Timeline (accessed July 17, 2019).

53    Cristina Maza, "Trump Administration Accusations That Russia Violated Nuclear Treaty Are a

'Cover Up,' Moscow Claims," *Newsweek*, June 17, 2019, https://www.newsweek.com/russia-trump-u-s-nuclear-treaty-tests-1444352.

54 John McCain, "Speech in Denver, 2008," *New York Times*, May 27, 2008, transcript of John McCain's University of Denver speech, https://www.nytimes.com/2008/05/27/us/politics/27text-mccain.html.

55 The White House, Office of the Press Secretary, "Fact Sheet: The Prague Nuclear Agenda," January 11, 2017, https://obamawhitehouse.archives.gov/the-press-office/2017/01/11/fact-sheet-prague-nuclear-agenda (accessed July 17, 2019).

56 US Senate, Committee on Foreign Relations, "New START Treaty: Resolution of Advice and Consent to Ratification," ratification document, 2011, https://www.foreign.senate.gov/imo/media/doc/SFRC%20New%20START%20Resolution%20FINAL.pdf.

57 Rachel Maddow, *Drift: The Unmooring of American Military Power* (New York: Broadway Books, 2012), 239.

58 Tom Collina, "Senate Approves New START," Arms Control Association, January 2, 2011, https://www.armscontrol.org/act/2011_01-02/NewSTART.

59 Interview with the authors, May 3, 2019.

60 Interview with Ben Rhodes, June 17, 2019, "Press the Button: Episode 10," podcast audio, *Press the Button*, Ploughshares Fund, edited for clarity with permission, https://soundcloud.com/user-954653529/ben-rhodes-former-deputy-national-security-advisor-joins-joe-cirincione-in-conversation.

61 Interview with the authors, May 3, 2019.

62 Interview with the authors, May 3, 2019.

63 Interview with Leah Greenberg, April 29, 2019, "Press the Button: Episode 03," podcast audio, *Press the Button*, Ploughshares Fund, edited for clarity with permission, https://soundcloud.com/user-954653529/leah-greenberg-indivisible-co-founder-one-of-times-100-most-influential-people-of-2019.

64 Abraham Lincoln quote can be found here: https://www.goodreads.com/quotes/489251-in-this-age-in-this-country-public-sentiment-is-everything.

## CHAPTER 10

1 Beatrice Fihn, "Nobel Lecture given by the Nobel Peace Prize Laureate 2017, ICAN, delivered by Beatrice Fihn and Setsuko Thurlow, Oslo, December 10, 2017," International Campaign to Abolish Nuclear Weapons, http://www.icanw.org/campaign-news/ican-receives-2017-nobel-peace-prize.

2 "Brown Returns to Public Eye, Issues Dire Warning at Doomsday Clock Announcement," CBS: San Francisco/Bay Area, January 24, 2019, https://sanfrancisco.cbslocal.com/2019/01/24/former-gov-jerry-brown-doomsday-clock-two-minutes-til-midnight.

3 *US Nuclear Excess: Understanding the Costs, Risks, and Alternatives: Executive Summary* (Washington, DC: Arms Control Association, April 2019), https://www.armscontrol.org/reports/2019/USnuclearexcess.

# INDEX

# INDEX

# INDEX

# ABOUT THE AUTHORS

**TOM Z. COLLINA** is director of policy at Ploughshares Fund. He has thirty years of Washington, DC, experience in nuclear weapons, missile defense, and nonproliferation issues, and has held senior positions at the Arms Control Association, the Institute for Science and International Security, and the Union of Concerned Scientists. He has been directly involved with efforts to end US nuclear testing, limit ineffective anti-missile programs, extend the Nuclear Nonproliferation Treaty, secure Senate ratification of the New START Treaty, and enact the Iran nuclear deal. He has published widely in major magazines and journals and has appeared frequently in the national media, including the *New York Times*, CNN, and NPR. He has testified before the Senate Foreign Relations Committee and regularly briefs congressional staff. Tom has a degree in International Relations from Cornell University and lives in Takoma Park, Maryland, with his wife, three children, and dog. When away from the office he does not think about nuclear war. This is his first book.

**WILLIAM J. PERRY** served as undersecretary of defense for research and engineering in the Carter administration and then as secretary of defense in the Clinton administration. He oversaw the development of the strategic nuclear systems that are currently in our arsenal. His new offset strategy ushered in the age of stealth, smart weapons, GPS, and technologies that changed the face of modern warfare. In 2007, Dr. Perry joined forces with George Shultz, Sam Nunn, and Henry Kissinger to publish several groundbreaking editorials in the *Wall Street Journal* that linked the vision of a world free from nuclear weapons with urgent but practical steps that could be taken to reduce nuclear dangers. Perry's 2015 memoir, *My Journey at the Nuclear Brink*, is a personal account of his lifelong effort to reduce nuclear dangers. He founded the William J. Perry Project to educate the public on these dangers, and is the Michael and Barbara Berberian Professor (emeritus) at Stanford University. Perry is the father of five, grandfather of eight, and great-grandfather of four. He continues to travel the world in pursuit of his goal of reducing the threat from nuclear weapons.